IARIA

CENTRIC 2019

The Twelfth International Conference on Advances in Human oriented and Personalized Mechanisms, Technologies, and Services

November 24 - 28, 2019

Valencia, Spain

CENTRIC 2019 Editors

Stephan Böhm, RheinMain University of Applied Sciences - Wiesbaden, Germany

Lasse Berntzen, University of South-Eastern Norway, Norway

Sandra Sendra, Universidad de Granada, Spain

Printed from e-media with permission by:

Curran Associates, Inc.
57 Morehouse Lane
Red Hook, NY 12571

CURRAN ASSOCIATES INC. proceedings .com

Some format issues inherent in the e-media version may also appear in this print version.

International Academy, Research, and Industry Association (IARIA)
412 Derby Way
Wilmington, DE 19810

Phone: (408) 893-6407
Fax: (408) 527-6351

petre@iaria.org

Additional copies of this publication are available from:

Curran Associates, Inc.
57 Morehouse Lane
Red Hook, NY 12571 USA
Phone: 845-758-0400
Fax: 845-758-2633
Email: curran@proceedings.com
Web: www.proceedings.com

Table of Contents

CENTRIC 2019

Forward

The Twelfth International Conference on Advances in Human-oriented and Personalized Mechanisms, Technologies, and Services (CENTRIC 2019), held on November 24 - 28, 2019- Valencia, Spain, addressed topics on human-oriented and personalized mechanisms, technologies, and services, commonly known as I-centric.

There is a cohort of technologies that favored the so called "user-centric" services and applications. While some of them reached some maturity, others are to prove their economics (WiMax, IPTV, RFID, etc). The human-oriented and personalized technologies and services rely on a key set of features, some to be deployed, others getting more mature (personal profiles, preferences, identity, proximity, personal devices, etc.). Following, advanced applications covering human related activities benefit from personalized and human-oriented networks and services, especially preventive and personalized medicine, body networks and devices, or anticipative systems.

The conference provided a forum where researchers were able to present recent research results and new research problems and directions related to them. The conference sought contributions presenting novel result and future research in all aspects of user-centric mechanisms, technologies, and services.

Similar to the previous editions, this event continued to be very competitive in its selection process and very well perceived by the international community. As such, it attracted excellent contributions and active participation from all over the world. We were very pleased to receive a large amount of top quality contributions.

We take here the opportunity to warmly thank all the members of the CENTRIC 2019 technical program committee as well as the numerous reviewers. The creation of such a broad and high quality conference program would not have been possible without their involvement. We also kindly thank all the authors that dedicated much of their time and efforts to contribute to the CENTRIC 2019. We truly believe that thanks to all these efforts, the final conference program consists of top quality contributions.

This event could also not have been a reality without the support of many individuals, organizations and sponsors. We also gratefully thank the members of the CENTRIC 2019 organizing committee for their help in handling the logistics and for their work that is making this professional meeting a success.

We hope the CENTRIC 2019 was a successful international forum for the exchange of ideas and results between academia and industry and to promote further progress in personalization research. We also hope Valencia provided a pleasant environment during the conference and everyone saved some time for exploring this beautiful city.

CENTRIC 2019 Steering Committee

Lasse Berntzen, University College of Southeast, Norway
Stephan Böhm, RheinMain University of Applied Sciences - Wiesbaden, Germany
Mu-Chun Su, National Central University, Taiwan
Nitin Agarwal, University of Arkansas at Little Rock, USA

CENTRIC 2019 Industry/Research Advisory Committee

Vitaly Klyuev, University of Aizu, Japan
Anbang Xu, IBM Research - Almaden, USA
Sabine Coquillart, INRIA, France
Thomas Marx, University of Applied Sciences Bingen, Germany

CENTRIC 2019 Publicity Chair

Sandra Sendra, Universidad de Granada, Spain

CENTRIC 2019

Committee

CENTRIC Steering Committee

Lasse Berntzen, University of South-Eastern Norway, Norway
Stephan Böhm, RheinMain University of Applied Sciences - Wiesbaden, Germany
Mu-Chun Su, National Central University, Taiwan
Nitin Agarwal, University of Arkansas at Little Rock, USA

CENTRIC Industry/Research Advisory Committee

Vitaly Klyuev, University of Aizu, Japan
Anbang Xu, IBM Research - Almaden, USA
Sabine Coquillart, INRIA, France
Thomas Marx, University of Applied Sciences Bingen, Germany

CENTRIC Publicity Chair

Sandra Sendra, Universidad de Granada, Spain

CENTRIC 2019 Technical Program Committee

Nitin Agarwal, University of Arkansas at Little Rock, USA
Mohammed A. Akour, Yarmouk University, Jordan
Mehdi Ammi, University of Paris-Saclay - CNRS/LIMSI, France
Alessio Angius, Polytechnic of Milan, Italy
Angeliki Antoniou, University of Peloponnese, Greece
Jose Luis Arciniegas Herrera, University of Cauca, Colombia
Aslihan Tece Bayrak, Media Design School, New Zealand
Lasse Berntzen, University of South-Eastern Norway, Norway
Stephan Böhm, RheinMain University of Applied Sciences - Wiesbaden, Germany
Marco Bruti, Telecom Italia Sparkle S.p.A., Rome, Italy
Susana Cadena Vela, Central University of Ecuador, Quito, Ecuador
Sonia Casillas Martín, University of Salamanca, Spain
Maher Chaouachi, McGill University, Canada
Daniel B.-W. Chen, Monash University, Australia
Sabine Coquillart, INRIA, France
José Luis Cuevas Ruíz, Instituto Federal de Telecomunicaciones, Mexico
Geeth De Mel, IBM Research, UK
Giorgio De Michelis, University of Milano - Bicocca, Italy
Yoanna Dosouto, Alaska Airlines, USA
Linda R. Elliott, Data and Analysis Center - U.S. Army Futures Command (AFC), Fort Benning, USA
Rainer Falk, Siemens AG, Germany
Tiantian Gao, Stony Brook University, USA
Francisco José García Peñalvo, Universidad de Salamanca, Spain
Alicia García-Holgado, University of Salamanca, Spain

Fabio Gasparetti, Roma Tre University, Italy
Nina Gerber, Karlsruhe Institute of Technology (KIT), Germany
Huan Gui, University of Illinois at Urbana-Champaign, USA
Riccardo Guidotti, KDD Lab | ISTI-CNR, Pisa, Italy
Till Halbach, Norwegian Computing Center, Norway
Qiang He, Swinburne University of Technology, Australia
Reiko Hishiyama, Waseda University, Japan
Jan Holub, Czech Technical University, Czech Republic
Tzung-Pei Hong, National University of Kaohsiung, Taiwan
Marius Rohde Johannessen, Universty of South-Eastern Norway, Norway
Imène Jraidi, University of Montreal, Canada
Christos Kalloniatis, University of the Aegean, Greece
Parves Kamal, St. Cloud State University, USA
Yasushi Kambayashi, NIT - Nippon Institute of Technology, Japan
Saman Kamran, University of Lugano, Switzerland
Teemu Kanstren, VTT, Finland
Vitaly Klyuev, University of Aizu, Japan
Boris Kovalerchuk, Central Washington University, USA
Jaroslaw Kozlak, AGH University of Science and Technology, Poland
Daniela Marghitu, Auburn University, USA
Patricia Martin-Rodilla, University of Santiago de Compostela, Spain
Célia Martinie, IRIT, University Toulouse 3 Paul Sabatier, France
Erik Massarczyk, University of Applied Sciences – RheinMain - Wiesbaden Rüsselsheim, Germany
Jose Manuel Molina Lopez, Universidad Carlos III de Madrid, Spain
Thomas Marx, University of Applied Sciences Bingen, Germany
Toshiro Minami, Kyushu Institute of Information Sciences, Japan
Sarah L. A. Morris, Cranfield University, UK
Alexios Mylonas, Bournemouth University, UK
Ana-Maria Olteteanu, Freie Universität Berlin, Germany
Yoshimasa Ohmoto, Kyoto University, Japan
Valentin Plenk, Hof University - Institute of Information Systems (iisys), Germany
Valentim Realinho, Polytechnic Institute of Portalegre, Portugal
Ann Reddipogu, RKode Ltd, UK
Michele Risi, University of Salerno, Italy
Elad Michael Schiller, Chalmers University of Technology, Göteborg, Sweden
Monica M. L. Sebillo, University of Salerno, Italy
Enric Senabre Hidalgo, IN3 - Internet Interdisciplinary Institute | Universitat Oberta de Catalunya, Spain / Austrian Centre for Digital Humanities (ACDH-ÖAW), Austria
Sandra Sendra, Universidad de Granada, Spain
Jingbo Shang, University of Illinois, Urbana-Champain, USA
Jungpil Shin, University of Aizu, Japan
Roman Shtykh, Yahoo Japan Corporation, Japan
Jacques Simonin, IMT Atlantique, France
Marie Sjölinder, RISE, Sweden
Alfredo Soeiro, University of Porto, Portugal
Christophe Strobbe, Hochschule der Medien, Stuttgart, Germany
Mu-Chun Su, National Central University, Taiwan
Stefan Sütterlin, Tallinn University of Technology, Estonia / Østfold University College, Norway

Jörg Thomaschewski, University of Applied Sciences Emden/Leer, Germany
Yuan Tian, University of Texas at Dallas, USA
Patricia Torrijos Fincias, University of Salamanca, Spain
Tsuda Toshitaka, Waseda University / Fujitsu Limited, Japan
Jacques Traoré, Orange Labs, France
Carlos Travieso González, University of Las Palmas de Gran Canaria, Spain
Filippo Vella, High Performance Computing and Networking Institute | National Research Council of Italy, Italy
Sanjaya Wijeratne, Kno.e.sis Center | Wright State University, USA
Anbang Xu, IBM Research - Almaden, USA
Sherali Zeadally, University of Kentucky, USA

Copyright Information

Enabling User Centered Distributed Stormwater Monitoring

Julio Aráuz

School of Emerging
Communication Technologies
Ohio University
Athens, Ohio 45701
Email: arauz@ohio.edu

Amy Lynch

Department of Geography
Ohio University
Athens, Ohio 45701
Email: lyncha@ohio.edu

***Abstract*—Stormwater management is a significant and expensive problem for cities and communities around the world. Impervious surfaces, such as roofs and parking lots, prevent water from being absorbed into the ground and result in large volumes of runoff. Stormwater runoff carries pollutants, such as oil and chemicals, across the urban landscape and into waterways and can lead to the overflow of combined sewer systems. Large volumes of stormwater can also lead to flooding and the erosion and sedimentation of waterways. Conventional systems of managing stormwater are centralized and expensive. Many communities have been moving away from centralized stormwater management systems and toward distributed solutions such as rain gardens. These strategies are designed to absorb urban runoff before it can leave the site. However, the distributed nature of these installations poses a major management challenge for cities and communities. There is a pressing need for low-cost monitoring solutions that enable a continuous understanding of the operation of these installations by their owners and by municipalities. In this work, we present an open source, open hardware design and first results of a sensing system that allows user centered monitoring of stormwater flows for rain gardens.**

Keywords*–*Urban; Stormwater; Management; User-centered.

I. Introduction

Over the past 20 years, American communities of all sizes have moved away from large, centralized storm water management systems and toward small-scale decentralized solutions. Decentralized strategies require a network of green infrastructure features such as rain gardens, swales, green roofs, and porous pavement which infiltrate and evaporate most of the runoff on-site, reduce the volume of runoff, and prevent the concentration of pollutants in waterways. Site-scale green infrastructure, also known as low impact development (LID) or site-based best management practices (BMPs), is designed to approximate the pre-development hydrology of a specific site [1]. Decentralized, green infrastructure, approaches can reduce the burden on existing centralized systems and delay or prevent the need for costly expansions or upgrades [2][3][4]. A combination of green and gray infrastructure projects can result in a lower total system cost than a centralized gray infrastructure system of tunnels, pipes, and treatment plants [2]. Green infrastructure also provides an array of ancillary environmental benefits and ecosystem services, such as wildlife habitat and reducing the urban heat island effect.

Despite a wide variety of benefits to this type of stormwater management, the distributed nature of the installations make monitoring a challenge and creates a barrier to broader adoption. Municipalities across the United States now require on-site stormwater management for new developments and provide financial incentives for landowners to retrofit existing properties with decentralized stormwater management. The result is tens of thousands of small-scale citizen-owned stormwater facilities distributed throughout the landscape. While these facilities may be small, each plays an important role. If an installation overflows or malfunctions it could lead to flooding, overloading of centralized stormwater systems, and the contamination of local waterways. Yet, it is difficult to know if these facilities are functioning properly. Stormwater managers must visually inspect each rain garden and green roof in the network. For local governments, who are accustomed to a few large centralized facilities, but are now depending upon the functionality of each component of a distributed network, this is a time-consuming and costly change. The monitoring challenge discourages further adoption of this promising strategy. As a consequence, there is a need to develop new means for understanding the functionality and efficacy of distributed stormwater facilities.

There is a pressing need for a technology-based approach to monitoring distributed management stormwater systems, particularly for rain gardens, which are among the most popular facilities. For these systems, where an area covered with especially chosen vegetation absorbs the collected stormwater, it is possible to design monitoring solutions by taking advantage of current low-cost sensing and wireless technologies. In such solutions, it is desirable to monitor the incoming water flow and ground humidity. The rate at which water is discharged into the garden can be used to understand if, and how well, the water collection system is operating. Ground humidity is also useful for estimating how much and how fast water can be absorbed by the garden. To monitor these variables, low complexity devices can be designed with a set of off-the shelf sensors, integrated with basic cellular communications boards. We designed, constructed, and evaluated a first version of such a system which currently allows the monitoring of stormwater flow and communicates real-time results to a cloud-based Internet connected data repository through a third generation wireless connection. All the components of the system were selected taking into consideration performance, accuracy and cost.

The management of distributed rain gardens and green roofs at the user level can provide municipalities with a granular understanding of the efficacy of these installations and enable them to create programs around user-centered incentives. These incentives can assume the form of credits towards already existing monthly sewage charges. In this manner, current and future owners of distributed green stormwater management infrastructure could be encouraged to install and properly maintain it and help municipalities with the mitigation of the effects of stormwater runoff.

Our work presented next starts in Section 2 with a discussion on the motivation supporting the use of distributed monitoring of stormwater flows. Thereafter, in Section 3 we visit relevant research literature in the area of green infrastructure. We then propose, in Sections 4 and 5, our design for a low-cost sensing device to monitor stormwater flows in rain gardens along an experimental design to evaluate its performance. In Section 6, we visit the the current limitations of our approach. Finally, in Section 7 we present our conclusions and discuss some viable possibilities of future work in the area.

II. MOTIVATION

Recent technological advances have allowed urban planners to think about infrastructure in new ways and catalyzed a "smart cities" movement. Smart city strategies use digital technologies to improve the efficiency and efficacy of city services and enhance the livability of urban areas. These technologies allow the remote monitoring of broadly distributed infrastructure like roads, bridges, and sewer systems.

While monitoring of all types of decentralized stormwater facilities would be useful, this study focuses on rain gardens, a prominent type of bio-retention. Bio-retention installations, such as rain gardens and grass swales, are vegetated depressed areas designed to receive, hold, and absorb stormwater. Research has confirmed the ability of bio-retention areas to reduce the volume and contamination of stormwater [5][6][7]. Rain gardens, in particular, are common in lower density urban and suburban areas where there is sufficient green space to support them. They can also be inexpensive, attractive, when well cared for, and require little investment or technical expertise to maintain. Many cities offer rain garden workshops or publish technical manuals to show landowners how to install and manage small bio-retention facilities. They are one of the more accessible strategies. But, while rain gardens are increasingly common, they must be carefully maintained to ensure that plantings survive and to manage erosion or sedimentation that could inhibit stormwater management functions. Poor maintenance can easily yield inadequate stormwater management. To ensure the overall stormwater system remains effective, local stormwater managers must be able to monitor the status of rain gardens.

There are currently no means to monitor distributed stormwater management systems implemented through rain gardens. While it should be possible to use sensing technology used to monitor centralized stormwater systems, this would certainly be cost prohibitive for a distributed solution. Consequently, there is a need and an opportunity to create a low-cost monitoring solution that facilitates management and enables the collection of real-time data.

III. RELATED WORK IN THE AREA

Current literature incorporates a variety of laboratory and field studies that have examined the ability of bioretention areas to reduce the volume and contamination of stormwater. Reviews of bioretention literature conclude that reductions in runoff volume and peak flow range between 40% and 90% are feasible [8]. A 2011 study found that a bioretention facility reduced the flow volume and rate from a parking lot by up to 99% [9]. Bioretention can also manage total suspended solids, bacteria, and temperature, if specifically designed to do so [7]. For example, average bacteria reduction for bioretention facilities with appropriate media ranges from 70 to 99% [8]. Studies also agree that bioretention is highly successful at removing metals, but identify variations in management of nutrients, particularly nitrate.

Early research using laboratory prototypes suggested that bioretention could reduce metal concentrations by over 90% and many nutrients by between 60 and 80% [10]. Nitrate was most problematic with 24%. Field studies have been similar. Results suggest that bioretention facilities are successful at removing nutrients and other pollutants, but the magnitude of those reductions varies. Researchers applied synthetic stormwater runoff to two bioretention facilities in Maryland, US and found removal of metals to be very high and nutrient management lagging. Total nitrogen and phosphorous loads declined by between 49 to 59% and 65 to 87%, respectively. A field study of six bioretention cells in North Carolina, US showed the facilities reduced metal and nitrogen loads, but were less successful with phosphorous [7].

The use of green infrastructure to capture stormwater and reduce the impact of combined sewer overflows has been recently studied for the case of mega-cities [11]. A mega-city is a defined as one with over 10 million residents. In particular, in the case of New York City, different green strategies, such as bioretention facilities, rain gardens and porous pavement, have been employed to reduce the amount of stormwater runoff. This mix of strategies resulted in reductions of stormwater runoff of up to 42% for the entire watershed and up to 55% for individual cases. Similar goals for large cities have also been studied for Chinese cities under the concept of "Sponge Cities". These cities also incorporate LID green infrastructure strategies. For the particular case of Beijing the current target is to capture up to 85% of annual precipitation through the strategies. Current modeling indicates that increases in the use of rain gardens, green roofs and permeable pavement will allow the city to achieve its capture target [12].

As discussed throughout this section, the benefits of green infrastructure to capture stormwater have been studied under varied circumstances. Nevertheless, the adoption of particular strategies at the residential level face some barriers. A particular incentive strategy could be the use of green infrastructure for urban agricultural use, which empirical studies have found to be a feasible option. The yield of vegetable rain gardens has been found to be comparable to that of a control traditional vegetable garden, while still reducing stormwater runoff by 90% [13].

The location of green infrastructure installations is also of interest to municipalities. This is particularly true for distributed rain gardens in individual residences. The selection of suitable sites is a key component of a city's strategy to maximize the benefits of runoff reduction. An approach that has been shown to be effective is to quantify the suitability of locations by analyzing soil attributes, slope, land use, ground water level, available area and cost using already existing geographical information systems [14].

While the use of green infrastructure to manage stormwater is being employed in numerous locations around the U.S., the monitoring of such distributed installations is still scarce [15]. One exception is an ongoing project launched in late 2016 by City Digital a public-private initiative in Chicago [16]. In this project, precipitation amounts, humidity levels, soil moisture measurements, air pressure, and chemical absorption rates are continuously monitored for large bio-swales. Then real-time data is collected available for management purposes. However, from the published details it is unclear what kind of sensing devices are employed or their cost.

IV. Low-cost Flow Rate Sensing Device for Rain Gardens

As discussed earlier, the use of distributed green infrastructure, such as rain gardens, can achieve significant reductions of stormwater runoff. However, the distributed nature of individual rain gardens pose a monitoring challenge, especially for small cities and communities with limited resources. Therefore, a main goal of our work is to develop a low-cost distributed monitoring solution that reports data over a wireless network to a centralized site.

Distributed monitoring devices can help municipalities understand the amount of rainwater captured and potentially identify problems with the installation. Such problems may occur when individual rain gardens stop collecting stormwater due to debris on collecting gutters. Therefore, we designed a device that can be directly attached to the lower part of stormwater collecting downspouts. Our device then measures stormwater activity right before the water is diverted towards the rain garden. We are not aware of any low-cost distributed solution similar to ours as currently rain gardens are typically installed without any monitoring.

A. *Variable Selection*

The performance level of a rain garden is a function of how well it can operate as a pervious system. There are two variables of interest we consider describe the overall performance of such a system. It is first necessary to understand how much water is flowing into the soil in the garden, thus it is vital to count with data that describes stormwater flow rate over time. Additionally, it is relevant to understand the humidity levels of the garden's soil. Under normal operation, the vegetation and the soil in the garden should regularly absorb the incoming stormwater. However, should the soil become saturated, normal operation is hindered and stormwater might overflow the garden.

B. *System Architecture*

The architecture of the system for monitoring the performance of a rain garden encompasses two main sensors, a flow rate sensor and a soil moisture sensor. In this work, we focus on the flow rate sensor, as affordable soil humidity sensors are readily available in the market. We designed a sensing device that can operate continuously as energy is collected through a small solar panel and stored in a battery. The design is illustrated in Figure 1, where stormwater at the input (U-01) passes through a custom flow rate sensor (U-04/05) towards the output (U-09) and then the rain garden (U-03). The output of the sensor is the deviation angle of the floating pendulum shown on the right of Figure 1. A microcontroller based device

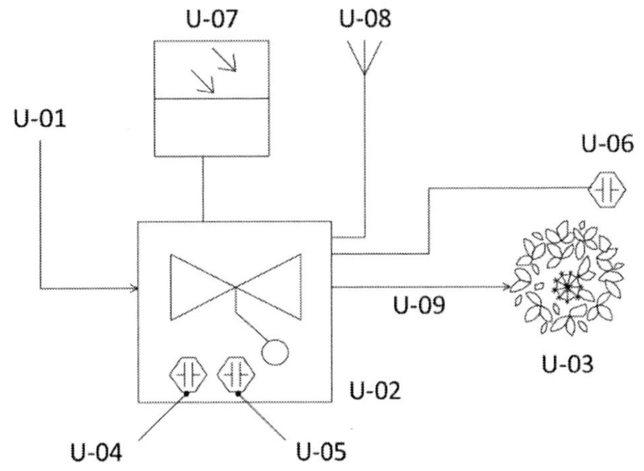

Figure 1. Architecture of the rain garden monitoring device.

(U-02), with onboard wireless cellular connectivity (U-08), controls the systems and collects telemetry data from the flow rate and a soil moisture sensor (U-06). The system is powered by a small 10W solar panel and battery (U-07).

1) Volumetric Flow Rate Sensor Design: While there are numerous commercial and industrial grade flow rate sensors, these are typically suited for measurements in clean water scenarios. These sensors are not suitable for stormwater measurements as they are prone to malfunction due to large contaminants or objects likely to be present in stormwater flows. Furthermore, these sensors are generally very expensive hindering the goal of constructing a low-cost device.

Using a simple approach based on a floating pendulum and low-cost fabrication methods we designed a first version of the flow-rate sensor. When stormwater flows through a pipe of inner diameter d, the sensor constantly measures the angle α between the pendulum axis and the vertical axis of the discharge pipe using an accelerometer. Figure 2 illustrates the geometry of the system. The flow rate pendulum sensor employs an accelerometer and gyroscope board based on the MPU-6050, micro electromechanical system (MEMS), manufactured by Invensense and available in small quantities for approximately 5 USD. When a stormwater flow is detected, this sensing board supplies the microcontroller with the angle α necessary to compute the flow rate. The board is installed in a custom designed, water sealed 3D printed housing that incorporates the floating pendulum. An exploded view of the housing and pendulum mechanism is shown in the top of Figure 3. The housing installed on a four inch plastic pipe is shown in Figure 4.

2) Sensor Calibration and Experimental Design: To estimate the volumetric water flow, Q, as a function of the angle α we calibrated the system using a typical setup consisting of a 2.5 meter tall downspout attached to a standard gutter. A known water flow rate was fed into the system and data was collected for eight different flow rates with three iterations of 2 minutes each. We heuristically set the data to be collected every 50 ms. This method has its limitations as it is dependent on the physical characteristics of a particular setup. Different installations will have downspouts of varying heights.

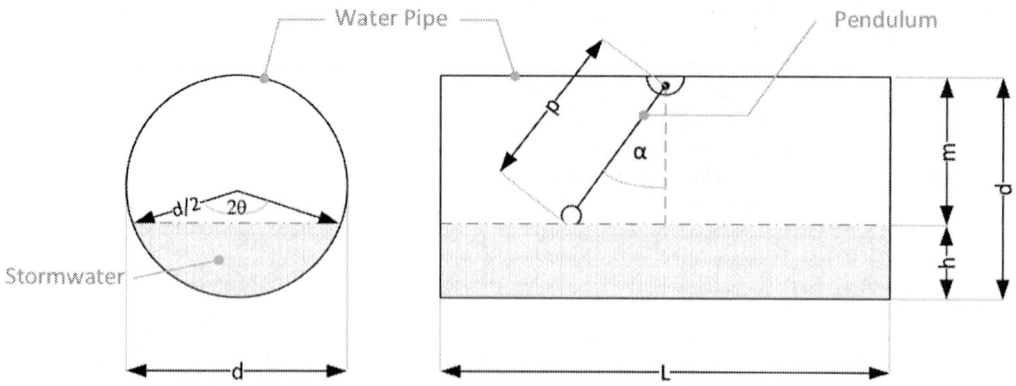

Figure 2. Flow rate sensor pendulum geometry when mounted on a 4 inch pipe (d = 4in., p = 3.2 in.) When stormwater flows through the pipe it rises to level h and deflects the pendulum an angle of α degrees. Front view (left), side view (right). Front view does not show the pendulum but just the stormwater level.

Figure 3. 3D exploded schematics of the hardware used to house the accelerometer board and the pendulum.

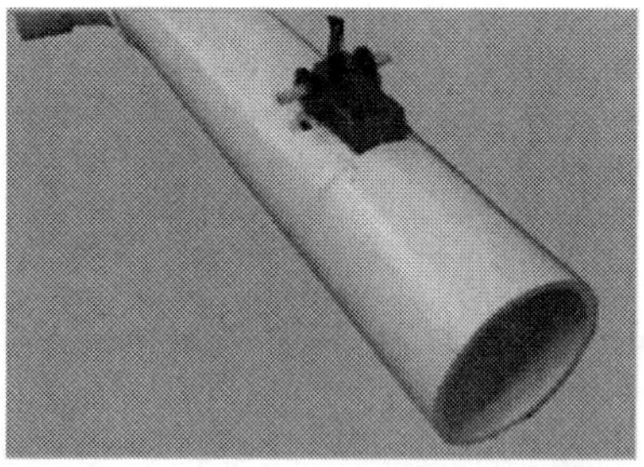

Figure 4. Hardware mounted on a 4 inch plastic pipe. The pendulum is not visible in the image as it lies inside the pipe.

However, the method can be extended to other scenarios as discussed in Section V.

3) Microcontroller and Wireless Connectivity: The design incorporates a microcontroller board to communicate with the MPU-6050, store real-time data, and provide cellular wireless connectivity. The Electron ARM based microcontroller board manufactured by Particle was chosen as it is a low cost, 65 USD, solution that incorporates 3G cellular connectivity as well. This microcontroller board requires very low power to operate drawing less the 1mA when no telemetry data is acquired, an average of 20 mA when the system is acquiring data and 800 mA during a two-minute duration batch wireless transmission. During wireless transmission of telemetry 15 minutes of previously recorded data is transmitted. The state of the pendulum (e.g., detecting water flow or not) dictates if data needs to be continuously collected every 50 ms or if the microcontroller should go to a low power state. This power saving behavior allows us to provide continuous 24 hour monitoring with a small 10 Watt solar panel.

The microcontroller board is currently configured to send information to two cloud-based data management platforms. It sends readings to Particle's dashboard reporting solution. Through the dashboard and custom code it also sends telemetry data to AT&T's M2X platform via the Message Queueing Telemetry Protocol (MQTT) protocol [17][18]. The cloud solution provided by AT&T allows data storage and real-time visualizations of the status of the system. It also enables data to be analyzed offline to carry out further studies of the results.

We selected MQTT as it being a tested telemetry protocol, software implementations are available for different platforms and programming languages. In MQTT, a broker receives published telemetry values from a client. Clients publish their data to a category referred to as topic. MQTT brokers can accept requests from other clients asking for this telemetry data to be shared. The simplest MQTT exchange is illustrated in Figure 5 where a client publishes data to the broker which in turn publishes it to an already subscribed client. MQTT provides for other types of exchanges that provide different guarantee levels to the messages being published.

All of our hardware designs, developed software, documentation, installation instructions and educational material is open and can be freely obtained online [19][20]. We expect this to greatly help not only in the reproducibility of our results, but towards the adoption of our design as a low-cost management solution by citizens and municipalities around the globe.

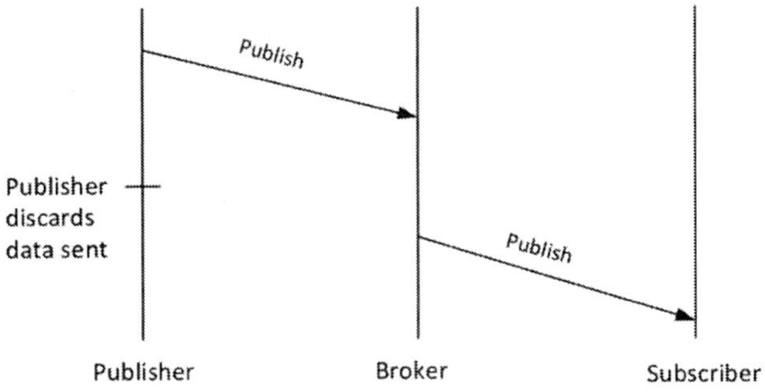

Figure 5. Sample publish of telemetry data via MQTT.

V. Flow Rate Sensor Results

The system we constructed was first calibrated using the procedure described in Section IV-B2, using eight input volumetric flow rates ranging from 0.0681 to 0.3419 liters per second. The resulting variations over time of the pendulum's deviation angle α for three of these levels are plotted in the top of Figure 6. Their respective histogram are shown in the bottom of the figure. The plotted variations over time are the result of applying a 20 sample average moving window (which corresponds to one second of samples collected every 50 ms).

The changes of the angle, α, as a function of time plotted in the top of Figure 6 ocassionally show significant variations around its mean value. While during the tests the flow rate stays relatively constant, the sampled angle varies around a mean value as the water flow is not laminar but turbulent. The turbulent nature of the flow was increased by the characteristics of our test setup where water rushing down the downspout collides with a 90 degree elbow connector before entering the sensor housing.

Nevertheless, when these variations are studied by plotting their histogram a unimodal distribution of the sampled angles was observed. These distributions are shown in the bottom part of Figure 6. This indicates that, under the conditions tested, the mean of the sampled angles is an adequate metric that can be used to estimate the flow rate.

The effect of all eight volumetric flow rates on the pendulum's angle is shown in Figure 7. Notice that the pendulum's angle for increasing flow rates should monotonically increase with the flow rate. However, during the experiments some increases in input flow rate resulted in decreasing values of the average angle (e.g., 0.1637 and 0.2310 liters per second) as evidenced by the data plotted in Figure 7. This is counterintuitive, as higher flow rate must result in larger pendulum deviations. After some investigation, we traced back this issue to random deviations of the actual input flow rate due to imperfections of the water pumping mechanism.

To be able to estimate flow rate for any angle we constructed a linear regression from our experimental data. The regression result is shown in Figure 7. The value of the coefficient of determination, R^2, for the regression was 0.91 which suggests our dataset approximately results in a linear relationship between the measured angle and a known input flow rate. A set of summarized statistics is provided in Table I.

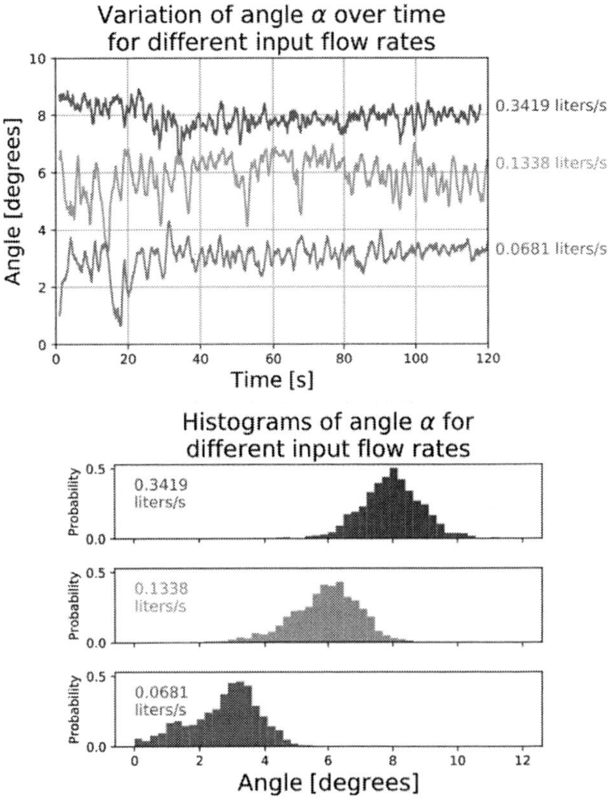

Figure 6. Variation and histogram of angle α for three different volumetric flow rates as measured by the accelerometer board.

TABLE I. SUMMARY OF EXPERIMENTAL DATA

Flow (l/s)	Mean α (degrees)	Standard deviation	Minimum (l/s)	Maximum (l/s)
0.3419	7.9440	0.9767	0.2259	11.7575
0.2310	4.7825	1.4651	0.5081	11.7580
0.1637	5.3209	0.9885	0.4581	8.4047
0.1338	5.8872	1.0612	1.7901	8.6158
0.1067	4.7872	1.0666	1.2105	8.4521
0.0876	3.5529	0.8082	1.1049	8.2843
0.0739	2.7729	0.6936	0.7093	6.2919
0.0681	2.6969	1.1247	0.7101	6.2919

VI. Limitations

It is possible to use the linear regression provided in Figure 7 to estimate the flow rate of stormwater passing through the sensor for any particular measured angle of the pendulum. However, this regression assumes the conditions the system is operating in are similar to those described in Section IV-B2. In different types of deployments, these assumptions might not hold and it may be necessary to estimate the flow velocity inside the sensor.

Given that stormwater generally flows inside a pipe following an *open channel* model where there is a pocket of air occupying part of the cross section of the pipe, estimating this velocity is not a straightforward process. Ideally our design could incorporate a flow velocity sensor; unfortunately this would increase maintenance complexity and price. Therefore,

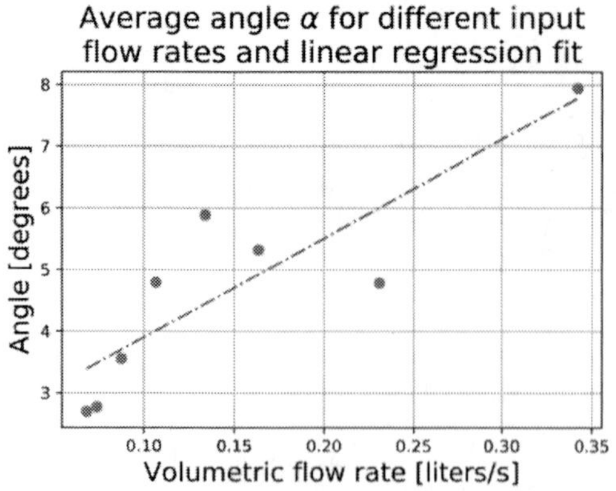

Figure 7. Variation of the average angle α for all eight input volumetric flow rates along with results from a linear regression.

we are currently using a different approach to estimate flow rate given a known deflection angle of the pendulum.

The method we use also employs a linear regression but uses as data the approximate flow rates given by the amount of rain water collected by the roof to which the sensor is attached. For any particular installation we can accurately compute the roof's surface area. The amount of rain water collected over time can be estimated by multiplying the surface area of the roof by the inches of rainfall reported in the area by weather services. The flow rate is then obtained by dividing the result by the length of the observation period. This method only provides us with approximation, but is commonly used in the sizing of residential and commercial stormwater downspouts and rain water collection systems.

While our approach is easy to implement it requires rainfall data that is only available from weather reports. On the other hand, this is a one time process. Once we have a linear regression to estimate flow rate through our device.

VII. Conclusion and Future Work

The first version of our Internet connected stormwater management device offers a robust solution at a low-cost. We estimate that in low quantities it is possible to construct the device for less than 120 USD. The current version does have limitations that need to be addressed, such as the need for calibration at a per-site level. This can be avoided by adding an additional sensor (e.g., flow velocity sensor). However, increasing the complexity would result in a higher cost. Thus we preferred to use an approximate per-site calibration procedure.

We have carried out preliminary tests of our open solution for a period of 30 days at two different rain gardens owned by the City of Athens, OH, with encouraging results. We are currently improving the design of our solution to allow the sensor to be mounted on different types of discharge pipes as well as to be used in green roof installations where both input and output flow volumes are of interest.

Acknowledgements

This work was partially supported by a grant from Ohio's Environmental Protection Agency Environmental Education Fund. The prototype of the sensing system was partially designed and constructed by Mr. Tristan Mull a Civil Engineering student with support from an Honors Tutorial College Research Apprenticeship Grant from Ohio University.

References

[1] US Environmental Protection Agency. *EPA: Green Infrastructure*. [Online]. Available from: https://www.epa.gov/green-infrastructure 2019.10.1

[2] H. Thurston, H. Goddard, D. Szlar, and B. Lemberg, "Controlling Storm-Water Runoff with Tradable Allowances for Impervious Surfaces," Journal of Water Resources Planning and Management, vol. 129, no. 5, 2003, pp. 409–418.

[3] E. Villareal, A. Semadeni-Davies, and L. Bengtsson, "Inner City Stormwater Control Using a Combination of Best Management Practices," Ecological Engineering, vol. 22, 2004, pp. 279–298.

[4] M. Keeley, "Using Individual Parcel Assessments to Improve Stormwater Management," Journal of the American Planning Association, vol. 73, no. 2, 2007, pp. 149–160.

[5] A. Davis, M. Shokouhian, H. Sharma, and C. Minami, "Laboratory Study of Biological Retention for Urban Stormwater Management," Water Environment Research, vol. 71, no. 1, 2001, pp. 5–14.

[6] D. M, "Low Impact Development Practives: A Review of Current Research and Recommendations for Future Directions," Water, Air, and Soil Pollution, vol. 186, 2007, pp. 351–352.

[7] W. Hunt and W. Lord, "Bioretention Performance, Design, Construction, and Maintenance," North Carolina State University Cooperative Extension, 2006.

[8] L. Ahiablame and B. Engel, "Effectiveness of Low Impact Development Practices: Literature Review and Suggestions for Future Research," Water Air Soil Pollution, vol. 223, 2012, pp. 4253–4273.

[9] K. DeBusk and K. Wynn, "Storm-water bioretention for runoff quality and quantity mitigation," Journal of Environmental Engineering, vol. 137, 2011, pp. 800–808.

[10] A. Davis, M. Shokouhian, H. Sharma, and C. Minami, "Water quality improvement through bioretention media: nitrogen and phosphorus removal," Water Environment Research, vol. 78, 2006, pp. 284–293.

[11] T. Eaton, "Approach and Case-Study of Green Infrastructure Screening Analysis for Urban Stormwater Control," Journal of Environmental Management, vol. 209, 2018, pp. 495–504.

[12] M. Randall, F. Sun, Y. Zhang, and M. Bergen-Jensen, "Evaluating Sponge City Volume Capture Ratio at the Catchment Scale Using SWMM," Journal of Environmental Management, vol. 246, 2019, pp. 745–757.

[13] P. Richards, C. Farrell, M. Tom, N. Williams, and T. Fletcher, "Vegetable Raingardens can Produce Food and Reduce Stormwater Runoff," Urban Forestry and Urban Greening, vol. 14, no. 3, 2015, pp. 646–654.

[14] V. Vaculova and J. Fuska, "Rain gardens – case study of potential locations identification using gis," Acta Scientiarum Polonorum. Formatio Circumiectus, no. 3, 2017, p. 217.

[15] US Environmental Protection Agency. *EPA: Green Infrastructure Cost-Benefit Resources*. [Online]. Available from: https://www.epa.gov/green-infrastructure/green-infrastructure-cost-benefit-resources 2019.10.1

[16] City Tech. *City Tech: City Solutions*. [Online]. Available from: https://www.citytech.org/city-solutions 2019.10.1

[17] AT&T. *AT&T: M2X API Documentation*. [Online]. Available from: https://m2x.att.com/developer/documentation/v2/overview 2019.10.1

[18] Message Queueing Telemetry Transport Organization. *MQTT: MQTT Version 5.0*. [Online]. Available from: https://docs.oasis-open.org/mqtt/mqtt/v5.0/mqtt-v5.0.html 2019.10.1

[19] Ohio University Smart Infrastructure. *Ohio University Distributed Stormwater Management*. [Online]. Available from: http://www.its.ohio.edu/raingardens 2019.10.1

[20] Ohio University Smart Infrastructure. *Distributed Stormwater Management Open Project*. [Online]. Available from: https://github.com/OUSmartInfrastructure/raingardenSensor 2019.10.1

Infrastructure-as-Code for Scientific Computing Environments

Daniel Adorno Gomes

University of Trás-os-Montes and Alto Douro
Vila Real, Portugal
www.utad.pt
Email: adornogomes@gmail.com

Pedro Mestre
and Carlos Serôdio

Centro Algoritmi
University of Minho, Guimarães, Portugal
CITAB - Centre for the Research and Technology of
Agro-Environmental and Biological Sciences
University of Trás-os-Montes e Alto Douro
www.utad.pt
Email: pmestre@utad.pt, cserodio@utad.pt

Abstract—**Infrastructure as Code (IaC) is receiving a lot of attention because of the positive results demonstrated in environment provisioning that supports both corporate and scientific applications. One of the biggest challenges related to the software development that supports scientific research projects is to obtain similar results to those that were published when that research is reproduced. Obtaining the same results involves recreating the same computational environment as the one used by the original researchers, and IaC is a paradigm that can help on this issue. In this paper, the authors provide a state-of-the-art and a literature review on IaC. Some work done on infrastructure provisioning based on IaC is presented and analyzed. A section shows how IaC can solve some typical issues on scientific computational environments. In other words, it is described how IaC practices were applied in a real case. Based on what is presented in this paper, and considering the scientific context, the use of IaC can help in many aspects like the quality of the developed software and the improvement of the results obtained in reproducible research.**

Keywords–IaC; devops; infrastructure-as-code; configuration script; continuous deployment; continuous delivery; continuous integration; reproducible research.

I. INTRODUCTION

The provision of the computational environment that supports scientific research has always been one of the main problems faced by researchers. In 2009, Jon Claerbout, a researcher at Stanford University who worked with data analysis and development of algorithms for geophysical exploration, reported that he had great difficulty reproducing algorithms and results reported by other researchers [1]. He reported also, that the same situation was occurring in his laboratory. Quite frequently his students and researchers faced problems to reproduce the results generated by their own research [2].

Even when we have similar setups, but on different servers, if for example the operating system is not the same, the software that will run on those servers can have different behaviors. It is therefore essential to have the same dependencies between the software components of the environment, to get similar results to those that have been disclosed. The computational environment used in the context of a particular scientific research has to be exactly the same in all aspects related to software, especially with respect to operating systems, compilers, libraries and their respective versions [3].

Enterprise and web applications differ in many aspects from scientific software, but when the subject is the infrastructure, they have faced the same issues. To guarantee the behavior of the applications will be the same on development, test and production environments, many companies like Github, Mozilla, Facebook, Google and Netflix have adopted an approach called Infrastructure-as-Code (IaC). The idea behind IaC is to provide the entire computer and network infrastructure through scripts. That is, no task that is related to the provisioning of the computational environment must be done manually, but in a programmatic way, both to avoid mistakes and to make the environment reproducible as many times as needed [4][5].

As we can see in Figure 1, there was a growing interest on IaC, mainly, in the last 6 years. The chart, obtained from Google Trends, shows the numbers of search results for the term "Infrastructure as Code" over the last 10 years.

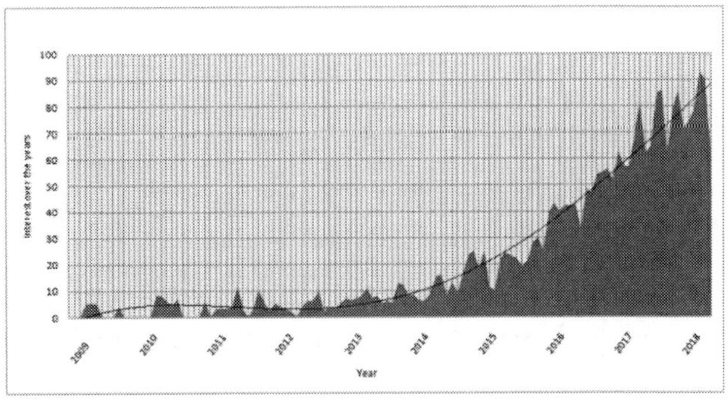

Figure 1. Search results for the term "Infrastructure as Code". Based on [4].

This paper presents a state-of-the-art and a literature review on IaC, that supports the section that highlights which IaC methods and techniques fit best for scientific computing environments.

The rest of the article is organized as follows: Section 2 presents the state-of-the-art on IaC. Section 3 makes a review of the literature on IaC. Section 4 presents a series of suggestions on how IaC can be applied on scientific computing environments. Section 5 describes a real case of applying IaC practices. Finally, in Section 6 it is presented the discussion and the conclusion.

II. Infrastructure-as-Code

In the last decade, due the growing of cloud computing demands, a new practice has been adopted to provide computing and network infrastructure through the use of source code. This approach, called Infrastructure-as-Code (IaC), permits to manage infrastructure the same way as if it was a software systems [6].

With IaC we have a dynamic infrastructure where the servers are created with all the configuration and software needed, just by using commands in executable files such as shell-scripts. All the operations like to create a new server, change its configuration, install or uninstall software depends on just running a script.

Using scripts to define our infrastructure means that our environments will have more consistency and it will be more reliable. Manual provisioning can have different interpretations of the same instructions, resulting in different configurations and faults that are not easy to identify, mainly, when these environments are not monitored. Scriptable infrastructure can guarantee a consistent monitoring, as well [7].

This approach is based on some practices as follows [8]:

- Definition Files: all configurations are defined in shell-script files called definition files. This means that all the updates that we need to apply in the infrastructure are executed from these files for example install a database, increase the memory of a Virtual Machine and create a new server. There is no manual intervention. As all the updates are defined by source code in scripts, the changes can be applied in test environments as much as necessary to certify the production environments will be modified correctly.
- Self-documented systems and processes: Instead of creating documents with instructions to be executed by humans, scripts can contain the documentation of the systems and processes and the commands that need to be applied in the infrastructure. Besides, they are more precise and consistently than humans when executing these instructions.
- Version control: all the code need to be kept in a version control system like Git or Github. This way the infrastructure can be versioned. Every configuration and every change is kept to help diagnose problems that might happen.
- Continuous test: like we have seen recently on modern software system, tests are essential to rapidly find errors, even in infrastructure source code. Like any other kind of software, we can set up continuous integration pipelines to test and guarantee the quality of our code. The continuous integration will support the continuous deployment and delivery practices.
- Continuous deployment: an update with many simultaneous changes on the infrastructure can present a high number of issues. Continuous deployment pipelines can help to mitigate this problem, once making small updates it is faster to find errors and fix them.
- Continuous delivery: with this practice it is possible to decrease the downtime of the systems on upgrades or fixes, using techniques like Blue-Green Deployment and Parallel Change to apply small updates.

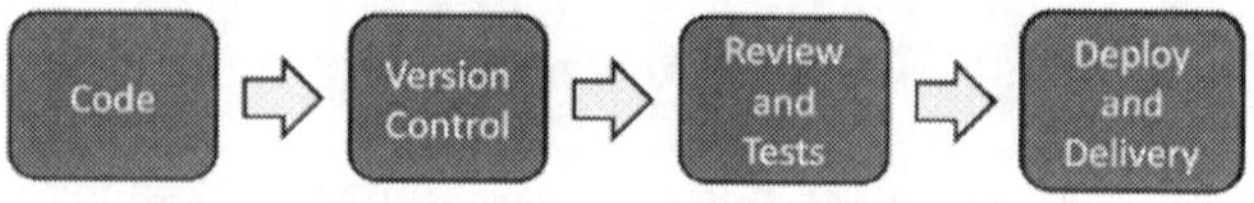

Figure 2. Infrastructure-as-Code workflow.

Adopting these practices, we can follow a well-defined workflow to provide infrastructure using IaC, as shown in Figure 2.

III. Related Works

In this section authors present a literature review and related works, made by other authors done in the field of infrastructure-as-code. These authors applied it to different areas of the software industry and scientific applications.

Garcia and Castillo [9] made a presentation about the main characteristics of the Cloud computing and virtualization techniques, discussing how benefit these features could be when used in scientific applications. In this paper, these authors highlight that neither virtualization nor cloud resources are commonly used in scientific computing environments due an idea that virtualization techniques have a negative impact on this kind of environments. They discuss the viability of the Infrastructure-as-a-Software cloud paradigm to attend the requirements of the computational science and the main issues that need to be addressed by the cloud players to provide the needed conditions to obtain the maximum benefit from this type of infrastructure.

Cole and Moore present in [10] a guide for creating biomedical workflows based on cloud computing environments. They show some cloud computing tools and characteristics that can help to increase reproducibility, scalability, resilience, fault-tolerance, security of software applications. Also they highlight how this paradigm can be cost and time efficient. They provide also an overview on how researchers can make the transition of their traditional biomedical informatics workflows to cloud environments.

In [11], Parnin et al. present a research work about the use of IaC in continuous deployment. Professionals from 10 different companies were interviewed by the authors. They reported the practice of IaC have changed the way how IT companies are managing their infrastructure. They also reported that the most frequent automation is related with unit testing, staging, and branching. When development focuses on delivery speed is essential to adopt continuous deployment, but some aspects like architecture and safety can have a decrease in quality, and they need a special attention.

In [12] members of the UC Berkeley D-Lab, Statistical Computing Facility (SCF), and Berkeley Research Computing (BRC) present strategies to reduce the complexity on creating scientific computing environments based on DevOps concepts. They start by showing some training, research use-cases and tools that support their strategies and help to create environments with more accessibility, productivity, reuse, and reproducibility. After, they present a DevOps framework that supported the creation of the Berkeley Common Environment (BCE). The BCE provides a standard reference that explains how to run a variety of projects on different platforms. This environment is available to be deployed on VirtualBox and Amazon EC2 environments, but it can be also provisioned for other platforms. It is expected to achieve identical results across any platform.

Howe [13] makes a presentation about the main characteristics of the virtualization on cloud computing platforms, focusing the use of virtual machines to provide computational environments for scientific research projects. He refers that a virtual machine can contain the entire working environment of the research project, including its data, all software, notes, logs and scripts, facilitating the reproducibility of the research. He presents a discussion about the consequences and benefits of the use of such a model, and shows the adaptations that are needed to use this approach in scenarios where of the reproducibility is more complex.

Boettiger presents in [14] the issues related with the reproducibility of the source code developed for a specific research project. He points the reasons why the code cannot be executed or extended by other researchers with success. Also, he provides a review on different approaches like virtual machines and workflow systems, showing their limitations. The Docker technology is analyzed by the author, that shows the advantages of the containerization like portability, reusability, versioning and cross-platform, and how it can address those challenges that are related to the provisioning of computational environments for scientific research.

IV. IaC for Scientific Computing Environments

Based on the state-of-the-art and the review of the literature presented in the above section, in this topic it will be highlighted the advantages of the adoption of the IaC practices and techniques, on scientific environment provisioning. Some typical issues faced by researchers will be shown and it will be discussed how IaC can help to solve them.

As mentioned in the above Introduction section, one of the most common issues reported by researchers on reproducing results is related with the computational environment. Less than 50% of software can be built or installed successfully. It is necessary for a huge effort to recreate the original environment and get similar results [15]. IaC fits perfectly in this scenario. Once the environment is defined in shell-script files and these files are stored in version control repositories, researchers can reproduce exactly the original environment anytime. Besides, the environment can be versioned and the results of the research can be related to each version.

Another issue reported is related to the documentation, and on how to install and run the code associated with a published work. This problem is related not just with the software environment but with the code produced by the researchers as well. Gilbert et al. [16] published a study where just 30% of the analyzed researches had their results reproduced due imprecise documentation. This kind of problem can be easily addressed using IaC practices once all the documentation is embedded in the definition files combining instructions and code in the same place.

A problem known as "code rot" is a kind of issue that affects the results of the reproducible research due updates of the computational environments. Before to start creating a new system, developers have to decide which will be the Operating System, the development language, the libraries and the database technology that will be used in the project. Practically, the new software will have many dependencies of this environment that are not static [14]. A simple update to fix bugs on the operating system or replace deprecated features in the libraries can change the results produced by the code. To mitigate this kind of concern, there are some practices used in IaC like containerization and virtualization that permit to create and store an image of the entire environment. In this way, the results can be reproduced by other researchers using the same infrastructure created by the original researchers [17].

V. A Case Study

In this section will be described, in a brief way, all the steps and resources needed to provisioning an environment to support a real scientific experiment on computer vision. This environment was used for the development of a software application capable of identifying canine dysplasia cases, from the analysis of x-ray images. This research involved researchers from the Agrarian and Veterinary Sciences School and the Science and Technology School of the University of Trás-os-Montes and Alto Douro, Portugal.

Initially, it was defined that the environment should be provisioned as a virtual machine. In this way, it could be recreated at any time, by any of the researchers, in any available platform, more specifically, stand-alone, on-premise or cloud platform. For this project it was decided empirically that the VM should have 4 GB of RAM and 50 GB of disk space. In Table I are shown which software applications were defined to compose the environment.

TABLE I. List of software applications defined to compose the environment.

Type	*Software*	*Version*
Operating system	Linux Ubuntu Server	Xenial v16.04 (LTS)
Programming language	Python	v2.7.12
Container platform	Docker	v18.09.0
Deep learning framework	Tensorflow	v1.7

The flow to provide this environment started with a tool called Vagrant v2.2.5. All the software and hardware needed to create the environment were described on Vagrant that generated a script with all these definition, and executed it creating a VM. The Figure 3 shows part of the script source code.

The first version of the script was uploaded to Github repository. The VM were used to execute tests on the installed software and their respective versions. After the test, the Vagrant script was updated with new commands to remove all versions of Python different of the version chosen by the researchers to avoid conflicts. After more tests, a new version of the script was uploaded to Github.

The last step was to validate if the code produced by the researchers could be compiled and ran properly on this environment, generating consistent results. Having a tested version of the script on Github, all the researchers could download it and create the same environment on their own equipment to reproduce the experiment, getting the same results.

```
  config.vm.provider "virtualbox" do |vb|
    # Display the VirtualBox GUI when booting the machine
    vb.gui = true
    # Customize the amount of memory on the VM:
    vb.memory = "4096"
  end

  # Installing and removing software applications
  config.vm.provision "shell", inline: <<-SHELL
    apt-get update
    apt-get install -y python2.7
    apt-get install -y docker.io
    apt-get remove -y python2.6

  SHELL
end
```

Figure 3. Script with the definitions of the environment.

This is a very simple example case study, but that can be used to show that the use of Infrastructure-as-Code can be very helpful to obtain reproducible research.

VI. CONCLUSION AND FUTURE WORK

In this paper the authors made a presentation of the state-of-the-art and a literature review on Infrastructure-as-Code and how it can be used to obtain reproducible research. Some aspects related to practices and techniques that are used by other researchers were discussed, and a workflow based on these items was proposed.

In the section about the work done by in this field, we could follow how IaC can have a very positive impact in software development development, both for enterprise and scientific applications. Also, this work discussed some typical issues related to scientific computational environments and showed how IaC can help to address them.

As future work, we propose to implement a scientific computational environment applying the practices of IaC, and compare it with traditional methods of computer infrastructure provision, in terms of costs and time.

The above presented methods will also be very helpful in other Data Center applications, in Virtual Machine provisioning, by defining the base infrastructure for other scientific projects under development in the authors Research Centres and University, or even to define the basic infrastructure to be used by students in their projects.

ACKNOWLEDGMENT

This work is supported by National Funds by FCT - Portuguese Foundation for Science and Technology, under the project scopes UID/CEC/00319/2019 and UID/AGR/04033/2010.

REFERENCES

[1] D. L. Donoho, A. Maleki, I. U. Rahman, M. Shahram, and V. Stodden, "Reproducible Research in Computational Harmonic Analysis," Computing in Science Engineering, vol. 11, no. 1, Jan 2009, pp. 8–18.

[2] M. Schwab, N. Karrenbach, and J. Claerbout, "Making scientific computations reproducible," Computing in Science Engineering, vol. 2, no. 6, Nov 2000, pp. 61–67.

[3] L. A. Barba and G. K. Thiruvathukal, "Reproducible Research for Computing in Science Engineering," Computing in Science Engineering, vol. 19, no. 6, November 2017, pp. 85–87.

[4] A. Rahman, R. Mahdavi-Hezaveh, and L. Williams, "Where Are The Gaps? A Systematic Mapping Study of Infrastructure as Code Research," Information and Software Technology, vol. 18, April 2018, pp. 65–77.

[5] M. de Bayser, L. G. Azevedo, and R. Cerqueira, "ResearchOps: The case for DevOps in scientific applications," in 2015 IFIP/IEEE International Symposium on Integrated Network Management (IM), May 2015, pp. 1398–1404.

[6] D. Chapman. Introduction to DevOps on AWS. [Online]. Available: https://d1.awsstatic.com/whitepapers/AWS_DevOps.pdf [retrieved: September, 2019]

[7] Introduction to DevOps on AWS. [Online]. Available: https://d1.awsstatic.com/whitepapers/AWS_DevOps.pdf [retrieved: September, 20179]

[8] K. Morris, Infrastructure as Code: Managing Servers in the Cloud, 1st ed. O'Reilly Media, Inc., 2016.

[9] Á. L. García and E. Fernández-del-Castillo, "Analysis of Scientific Cloud Computing requirement," in Proceedings of the 7th Iberian Grid Infrastructure Conference, Madrid, Spain, September 2013, pp. 147–158.

[10] B. S. Cole and J. H. Moore, "Eleven quick tips for architecting biomedical informatics workflows with cloud computing," PLOS Computational Biology, vol. 14, no. 3, March 2018, pp. 1–11.

[11] C. Parnin, E. Helms, C. Atlee, H. Boughton, M. Ghattas, A. Glover, J. Holman, J. Micco, B. Murphy, T. Savor, M. Stumm, S. Whitaker, and L. Williams, "The Top 10 Adages in Continuous Deployment," IEEE Software, vol. 34, no. 3, May-June 2017, pp. 86–95.

[12] D. Clark, A. Culich, B. Hamlin, and R. Lovett, "BCE: Berkeley's Common Scientific Compute Environment for Research and Education," in Proceedings of the 13th Python in Science Conference, Austin, USA, July 2014, pp. 5–12.

[13] B. Howe, "Virtual Appliances, Cloud Computing, and Reproducible Research," Computing in Science Engineering, vol. 14, no. 4, July-August 2012, pp. 36–41.

[14] C. Boettiger, "An Introduction to Docker for Reproducible Research," ACM SIGOPS Operating Systems Review - Special Issue on Repeatability and Sharing of Experimental Artifacts., vol. 49, no. 1, Jan. 2015, pp. 71–79.

[15] C. Collberg, T. Proebsting, G. Moraila, A. Shankaran, Z. Shi, and A. M. Warren. Measuring Reproducibility in Computer Systems Research, Technical Report. [Online]. Available: http://reproducibility.cs.arizona.edu/tr.pdf [retrieved: Septeber, 2014]

[16] K. J. Gilbert, R. L. Andrew, D. G. Bock, M. T. Franklin, N. C. Kane, J.-S. Moore, B. T. Moyers, S. Renaut, D. J. Rennison, T. Veen, and T. H. Vines, "Recommendations for utilizing and reporting population genetic analyses: the reproducibility of genetic clustering using the program STRUCTURE," Molecular Ecology, vol. 21, 2012, pp. 4925–4930.

[17] J. Ooms, "Possible Directions for Improving Dependency Versioning in R," The R Journal, vol. 5, 2013, pp. 197–206.

Difference in Attitudes Toward Suggestions Given by an Agent Using Impasse Estimation

Yoshimasa Ohmoto
Department of Intelligence Science and Technology
Graduate School of Informatics
Kyoto University
Kyoto, Japan
Email: ohmoto@i.kyoto-u.ac.jp

Hanako Sonobe
Department of Intelligence Science and Technology
Graduate School of Informatics
Kyoto University
Kyoto, Japan
Email: sonobe@ii.ist.i.kyoto-u.ac.jp

Toyoaki Nishida
Department of Intelligence Science and Technology
Graduate School of Informatics
Kyoto University
Kyoto, Japan
Email: nishida@i.kyoto-u.ac.jp

Abstract—**When a human and a conversational agent cooperate to perform a task, the agent may need to provide advice to the human. In that case, the agent has to induce attitudes that ensure the human accepts the advice of the agent in advance or through the task. This study aimed to investigate whether the acceptance of advice provided by an agent could be encouraged by controlling the content and timing of advice presentation according to the state of the participant. We focused on metacognitive suggestions during insight problem solving as an example of advice that was effective even if an agent performs it. We conducted an experiment to investigate whether participants would be likely to accept the suggestion based on the estimation of the inner state of them even when the agent provided the suggestion. Based on results from the analysis of operation history log, the acceptance rate of suggestions in the participants interacting with the state-considering-suggestion agent was significantly higher than that in the participants interacting with fixed-interval-suggestion agent, but the task performance was not high enough.**

Keywords–Human-agent interaction; Metacognitive suggestion; Insight problem solving

I. Introduction

When people work collaboratively with others, they often communicate their task status to each other and point out problems. This kind of information sharing and advice can be misunderstood or ignored unless people consider the state of the person (e.g., thinking, confusing, busy and so on) with whom they are communicating. For example, repeated advice about what is not a problem leads to disregard of the advice itself. These problems become more apparent when the collaborating partner is an artificial agent. On the other hand, the collaboration can facilitate problem-solving [1] [2], so we want to realize the agent which collaboratively supports problem-solving.

In order to avoid such a matter, the agent needs to understand the human state and give advice; however, even a human often fails to estimate the inner state of the other person. When you have a trusting relationship with the communication partner, it is often not a fatal problem. However, in the human-agent interaction, the human often needs to infer the agent's behavior model from the case of a few interactions. Therefore, it is expected that a small number of failures in interaction will cause errors in the behavioral model of the agent constructed by humans. For example, one approach to getting people to accept an agent's advice is to show that the agent has expertise by providing the appropriate advice [3]. The human often accepts the advice of the agent when the agent provides appropriate advices depending on the task situation. However, if it fails early in the interaction due to misunderstandings, it may stop accepting further advice. In addition, it is often hard to determine whether or not the advice is appropriate when it is not the advice that leads to the correct answer for the task being performed. In such instances, effective advice cannot be given without considering the situation and intention of the person performing the task.

Metacognitive suggestions are one of the useful suggestions for problem solving, although it does not lead to the direct outcome of the task being performed. Several previous studies had attempted to improve task performance by pre-training to induce meta-cognition. Patrick et al. [4] reported the impact of general meta-cognitive training on performance. Metacognitive suggestions may convey knowledge of how participants solve problems and can facilitate changes in their way of thinking during insight problem solving (e.g., [5]). In this study, we consider metacognitive suggestions during insight problem solving that is effective even if an agent performs it.

This study aimed to investigate whether the acceptance of metacognitive suggestions provided by an agent could be encouraged by controlling the content and timing of suggestion presentation according to the state of the participant. Basically, the timing of an agent's advice with participants occurs in a silence section of the conversation in many previous works (e.g., [6], [7]). To make the agent a more powerful assistant, content recognition was tried to be incorporate in many cases, but it is difficult. In this study, to provide acceptable agent's advice, we did not focus on the content of the solving problems, but on the thinking state and feeling of difficulty toward the solving problems. The advantage of this research approach is that it is possible to provide advice at an appropriate time based on the difficulty of the task that the person feels, without recognizing the content of the task or what the person is doing. If the effectiveness of an agent's advice depends on whether the agent can give advice based on the state of the person, it is expected that an influential suggestion can be given to a person without building a certain trust relationship with the agent through long-term interaction.

This paper adds details of the developed system, data of participants and the analyses and discussions that were presented as a Late Breaking paper at 6th international conference on Human-Agent Interaction [8].

The present paper is organized as follows. Section II contains an explanation of a system developed to give metacognitive suggestions based on the estimated state of the person performing the insight problem-solving task. Section III describes the results of an experiment to evaluate the system implemented on the agent. In Section IV, the achievements of this research and some future works are described. The conclusions are presented in Section V.

II. SUGGESTION SYSTEM USING IMPASSE ESTIMATION

Insight problem solving contains four steps: impasse, incubation, illumination, and validation [9]. We focus on the impasse step in which people repeatedly searches inappropriate problem space that does not include a solution. In the impasse step, advice from other perspectives is useful for constraint relaxation and a switch of problem space. Metacognitive suggestion is one method of providing acceptable advice for the constraint relaxation [1], [10], [11]. The metacognitive suggestion is confirmed to be effective even if it is presented at random timing. This is because the insight problem-solving task is prone to fall into the impasse state, and therefore there is a certain probability of being in the impasse state when presented at random.

In order to confirm the appropriate timing of advice in an insight problem-solving task, we conducted a preliminary experiment in which an experimenter determined the content and timing of the agent's metacognitive suggestions using the Wizard of Oz (WoZ) and presented them to the person performing the task. The task in the preliminary experiment was an "escape room game" in which players were often at a stalemate because they were required to think from a different perspective to win this game. In this game task, the participant must escape from a virtual room using various game objects. After this preliminary experiment, some participants reported that they were "given proper advice", so we thought that the advice of the agent by WoZ operation was accepted. When we observed the behavior of the participants and advice of the experimenter in the preliminary experiment, the experimenter provided suggestions when the participant seemed to be at a stalemate. We regarded this state of stalemate as an impasse. We consider that the state of stalemate is one of the appropriate clues to provide metacognitive suggestions. In accordance with this concept, we developed a system to provide advice by estimating whether the interaction partner is in an impasse state while working on an insight problem-solving task.

A. Estimation of strategies to perform the insight problem solving task

In order to find typical strategies to perform the insight problem-solving task, we observed the behavior of the participants in the preliminary experiment. As a result, it was expected that the participants switched two strategies: depth-first search and breadth-first search. In the state of depth-first search, participants focused on a particular object and looked for ways to use it successfully. In the state of breadth-first search, participants saw the overall situation of the task to

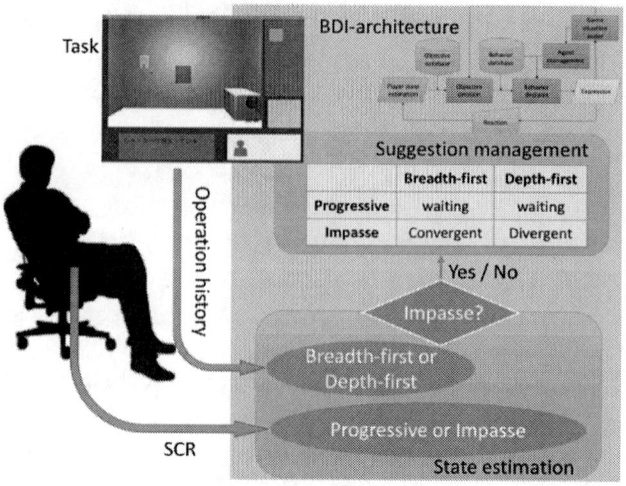

Figure 1. Outline of the agent design.

search whether there were any missing or untried methods. Since it is conceivable that a stalemate may occur while executing each strategy, human inner states in insight problem solving can be classified into 4 states (table in Figure 1). In the advice by the experimenter, there were many suggestions that urged the participant to look for other ways to solve the task in the depth-first search, and there were many suggestions that encouraged the participant to look back on his/her own behavior and to focus on the specific object in the breadth-first search.

It is difficult to infer the inner state of thinking from the participant's behavior, specifically the inner state of thinking whether the participant is at a stalemate. To estimate this inner state, we analyzed physiological indices obtained during the preliminary experiment. In our previous work, we reported to estimate the feeling of difficulty of the task by using physiological indices [12]. As a result, it was frequently observed that Skin Conductance Response (SCR) was often activated, when the unfamiliar object was discovered during the task and when the situation in the task was changed. In addition, even when the situation did not change, the SCR was often responsive when with repeated trial and error such as looking for hints or checking previous information. Therefore, we regard the state as a non-impasse state (the participant is not at a stalemate) when the responses of SCR are frequently observed, and we regard the state to have shifted to the impasse state (the participant is at a stalemate) when the response of SCR is not observed for a certain time.

We also measured the electrocardiogram. However, we have not been able to obtain a useful feature for estimating task impasse from the electrocardiogram. Therefore, no electrocardiogram data was input to the system. We used electrocardiographic data to assess participant's psychological status to tasks.

B. The outline of the suggestion system using impasse estimation

Figure 1 shows an outline of the agent design. This agent basically decides own behavior based on the Belief-Desire-

Intention (BDI) architecture. This agent estimates two kinds of user states: a thinking mode (depth-first or breadth-first) and a state of the stalemate (impasse or progressive). The user's overall states are categorized into one of four combinations: depth-first/progressive, breadth-first/progressive, depth-first/impasse, and breadth-first/impasse. The agent provides a metacognitive suggestion when the estimated user's state includes "impasse." A convergent suggestion is provided in the state of breadth-first/impasse. A divergent suggestion is provided in the state of depth-first/impasse.

The user's physiological index and behavior are measured to estimate the user's state. The state of the stalemate is estimated using the measured SCR. The agent estimates the user's state as impasse when the SCR does not respond during a defined time window. The time window and the threshold to estimate the state of the stalemate are decided based on the measured data for two minutes from the start of the task. To estimate the thinking mode, the operation history log of the user is used. When the user repeatedly operates a game object (such as, a key, a scissors, a piece of paper, a door, a dial plate, a drawer of a desk, a closet, a button and a safe) in high frequency, the agent estimates the thinking mode to be depth-first search.

Ten convergent and ten divergent suggestions were prepared. The suggestions were not dependent on a particular task because they were metacognitive suggestions. One of the suggestions is selected randomly when the agent provides an advice. In general situations, it is necessary to give advice considering the context of the task though it is a metacognitive suggestion. In this study, we focused on the effect of controlling the timing of the metacognitive suggestions provided based on the state of the user. Therefore, the agent advised only considering whether the context was divergent or convergent in our experiment.

III. Experiment

When we try to intervene in the behavior or decision-making of the other person by providing advice, especially during interaction with a less socially related interaction partner, it is important to provide appropriate advice based on the estimation of the partner's inner state. We considered the metacognitive suggestion in the insight problem-solving task as an example of the useful advice that the agent can provide, and proposed the suggestion system based on the impasse state estimation of the partner in order to accept the suggestion by the agent. We used two types of suggestion agents in this experiment. One was a state-considering agent that estimated the user's state before providing a metacognitive suggestion (sc-group). Another was a fixed-interval agent that provided a metacognitive suggestion in three-minute intervals (fi-group).

A. Task

Participants played an "escape from the room" game. In this game, players were often at a stalemate because they are required to think from a different angle to escape from a virtual room using various game objects. The participants are asked to escape from three rooms. In order to gradually increase the difficulty, the order of the rooms that the participant escaped from was fixed. There was a 15 minute time limit to escape from each room. The suggestion agent explained the procedure for escaping from the room when the participant exceeded this time limit. After escaping from a room, the participant was allowed to rest.

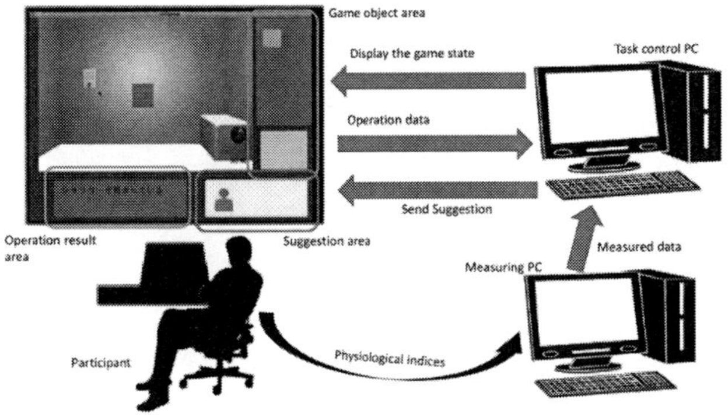

Figure 2. The experimental setting.

B. Exprimental setting

Figure 2 shows the experimental setting. Each participant sat in front of a 27-inch monitor that displayed the game. A video camera was placed behind the participant to record his/her behavior and the game playing screen. The participant's voice was recorded using microphones. Polymate was used to measure SCR and the electrocardiogram (heart rate variability). SCR was measured with electrodes attached to the first and third fingers of the participant's non-dominant hand. The electrocardiogram was measured by connecting electrodes with paste to the participant's left side, the center of the chest, and both ears for ground and reference. The experimenter sat out of view of the participant and observed the participant's behavior. The suggestions by the agent were provided using audio and text. The participants performed the task using a mouse.

C. Procedure

First, each participant was briefly instructed on the experimental procedure. Electrodes for measuring SCR and the LF/HF electrocardiogram values were then attached to the participant's left hand and chest. After the installation, each participant played a practice game to confirm the operating method and basic flow of the game. The experimenter instructed the participant on the basic operation method. In addition, the participant was given an overview of the agent providing metacognitive suggestions. After receiving questions from the participant and confirming his/her understanding, the participant started the "escape from the room" experiment. After the experiment, the participant answered NASA-TLX to measure the mental workload.

Forty-two undergraduate students, 27 males and 15 females, participated in the experiment. The average age was 20.8 years with a standard deviation of 1.9 years. We eliminated 13 participants because they did not need suggestions to escape from one of the rooms. Therefore, we used data of 29 participants (sc-group: 14 participants, fi-group: 15 participants). We conducted an additional experiment for five undergraduate students (three males and two females) from [8], but one male person did not come and three (two males and one female) were eliminated. Added one female participant was fi-group.

D. Results

1) The frequency of metacognitive suggestions : We analyzed whether there was a difference in the frequency of

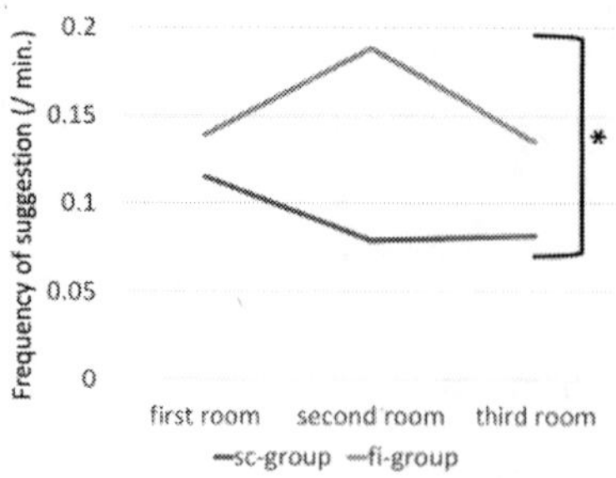

Figure 3. The frequency of divergent metacognitive suggestions per minute.

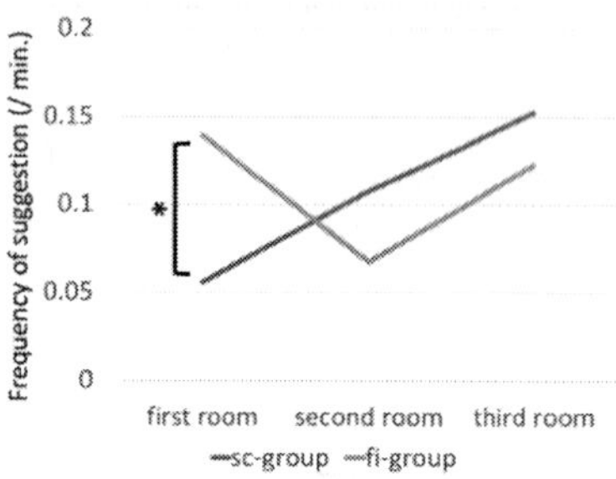

Figure 4. The frequency of convergent metacognitive suggestions per minute.

metacognitive suggestions provided in the sc-group compared with the fi-group. There were two types of the metacognitive suggestions (divergent and convergent), so we performed a 2 (group: state-considering or fixed-interval) x 3 (room: first, second or third) analysis of variance (ANOVA) separately. Since each participant spent different amounts of time in each room, we compared the number of suggestions per minute. Logit transformed values were used in ANOVA to test for differences. The results are shown in Figure 3, Figure 4, Table I and Table III.

In the divergent suggestions, there were significant differences between groups (sc-group < fi-group) and between rooms (first and second > third). The interaction was also significant. When tested for simple main effects, there were significant differences between groups in second room and third room (sc-group < fi-group). There was also a significant difference between the rooms in the sc-group (first > second and third). This result indicates that, in the sc-group, a relatively large amount of divergent suggestions was provided in the first room where the task execution method is unclear for the participants, and that trial and error is encouraged. In addition, in second room and third room where people seem to be used to the task, suggestions were reduced.

In the convergence suggestions, there was no significant difference between groups, but there were significant differences between rooms (first and second < third). The interaction was also significant. A simple main effect test showed a significant difference between groups in first room (sc-group < fi-group). It was also found that there were significant differences between the rooms in the sc-group (first < second and third). This result shows that the convergent suggestions in the sc-group was reduced in first room which was a relatively

TABLE I. RESULT OF THE ANALYSIS OF VARIANCE ON THE FREQUENCY OF DIVERGENT METACOGNITIVE SUGGESTIONS.

source	SS	df	MS	F	p
A: group	6.50	1	6.50	14.83	<0.001 ****
error[S(A)]	11.84	27	0.44		
B: room	1.81	2	0.90	5.13	0.0091 **
AB	1.17	2	0.58	3.32	0.044 *
error[BS(A)]	9.49	54	0.18		

TABLE II. THE SIMPLE MAIN EFFECT OF THE ANALYSIS OF VARIANCE ON THE FREQUENCY OF DIVERGENT METACOGNITIVE SUGGESTIONS.

effect	SS	df	MS	F	p
A(first)	0.38	1	0.38	1.43	0.235
A(second)	4.32	1	4.32	16.39	<0.001 ****
A(third)	2.98	1	2.98	11.30	0.0012 ***
error		81	0.26		
B(state-considering)	1.90	2	0.95	5.42	0.0072 **
B(fixed-interval)	1.07	2	0.53	0.04	0.056 +
error		54	0.18		

simple.

Overall, the control of metacognitive suggestions was reasonable to some extent.

2) Operation history log: After the metacognitive suggestion was provided, we analyzed from participants' operation history log to determine whether they were acting in line with the suggestion. From the operation history log, we checked whether the transition to another state occurred within 10 second after the suggestion was given. In divergent suggestions, if a state transition was made, it was considered that the suggestion was accepted. In convergent suggestions, if no state transition was made, the suggestion was accepted. The result is shown in Figure 5 and Figure 6. The chi-squared test was applied to determine whether there was a difference between the groups in the acceptance rate of divergent suggestions and the acceptance rate of convergent suggestions.

We compared the acceptance rate of all suggestions between groups. As a result, the acceptance rate of all suggestions in the sc-group was significantly higher than that in the fi-group ($p = 0.0013$). We compared the acceptance rates of divergent suggestions and convergent suggestions between groups. Although there was no significant difference in divergent suggestions ($p = 0.01$), the acceptance rate of convergent suggestions in the sc-group was significantly higher than that in the fi-group ($p = 0.0061$). We compared the acceptance rates of divergent suggestions and convergent suggestions in each group between rooms. In third room, the acceptance rates of both divergent suggestions and convergent suggestions in the sc-group were significantly higher than those in the fi-group (divergent: first room $p = 0.72$, second room $p = 0.16$, third room $p = 0.005$, convergent: first room $p = 0.44$, second room $p = 0.36$, third room $p = 0.014$). In addition, in the sc-group, the acceptance rates in third room were higher than those in first room, and the acceptance rates seems to be gradually increasing. It is not clear whether this is because the difficulty of the room is increasing or because the reliability of the agent's suggestions is increasing. In any case, the results showed that the participants were likely to accept the metacognitive suggestions provided by the agent when the suggestions were given based on the inner state estimation

TABLE III. RESULT OF THE ANALYSIS OF VARIANCE ON THE FREQUENCY OF CONVERGENT METACOGNITIVE SUGGESTIONS.

source	SS	df	MS	F	p
A: group	0.77	1	0.77	2.03	0.165
error[S(A)]	10.21	27	0.38		
B: room	1.15	2	0.57	4.10	0.022 *
AB	3.17	2	1.58	11.30	<0.001 ****
error[BS(A)]	7.57	54	0.14		

TABLE IV. THE SIMPLE MAIN EFFECT OF THE ANALYSIS OF VARIANCE ON THE FREQUENCY OF CONVERGENT METACOGNITIVE SUGGESTIONS.

effect	SS	df	MS	F	p
A(first)	3.83	1	3.83	17.42	<0.001 ****
A(second)	0.015	1	0.015	0.067	0.796
A(third)	0.099	1	0.099	0.45	0.50
error		81	0.26		
B(state-considering)	3.19	2	1.60	11.39	<0.001 ****
B(fixed-interval)	1.12	2	0.56	4.01	0.024 *
error		54	0.14		

of the participant.

3) LF/HF: LF/HF was calculated from heart rate variability data. LF/HF can be used as an index of tension and stress state of participants. The feeling of tension after the suggestions shows that it is influenced by something such as thinking after listening to the suggestions or feeling impatient. The rate of suggestions with LF/HF responses during the 30 s after the suggestion was calculated. Figure 7 shows the results. When the chi-squared test was applied to each room and to the whole data, the rate in the sc-group showed significantly higher than that in the fi-group in third room and the total in the task (third room: p=0.046, total: p=0.0060). There were marginal significant differences in first room and second room (first room: p=0.054, second room: p=0.050). This suggests that the metacognitive suggestions in the sc-group had a stronger effect on the participants than those in the fi-group.

4) Mental workload: We measured mental workload using NASA-TLX. This is major method to measure the mental workload. Figure 8 shows the results. With the exception of "performance," the sc-group reported an overall lower mental workload than the fi-group. We performed Welch's t-test on the total score between the two groups and there is no significant difference (p=0.17). We also performed Welch's t-test on each individual score between the sc-group and the fi-group. There was a significant difference regarding the data of "temporal demand" (sc-group < fi-group, p=0.038). The results suggest that advice based on human internal state estimation reduces some of the human mental workload. At the same time, it shows that the overall effect is not significant.

5) Task performance: We calculated success rates of escaping from the rooms. Figure 9 shows the results. In the third room, the participants in the sc-group could not escape from the room. The total success rate in the sc-group was lower than that in the fi-group. This was an unexpected result. Although the participants started to accept the suggestions by the agent, it is probable that the participants were confused and distanced themselves from solving the task because they were given suggestions that did not contribute to solving the task.

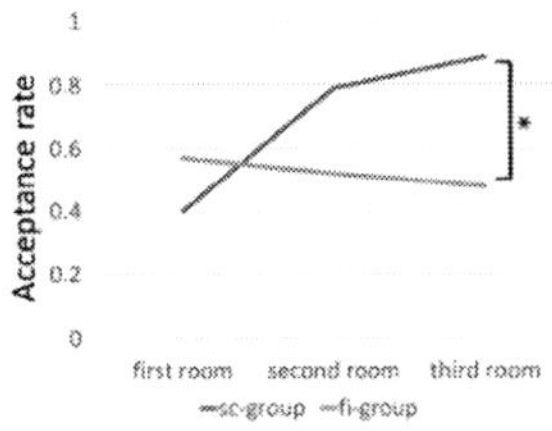

Figure 5. Acceptance rates of divergent metacognitive suggestions.

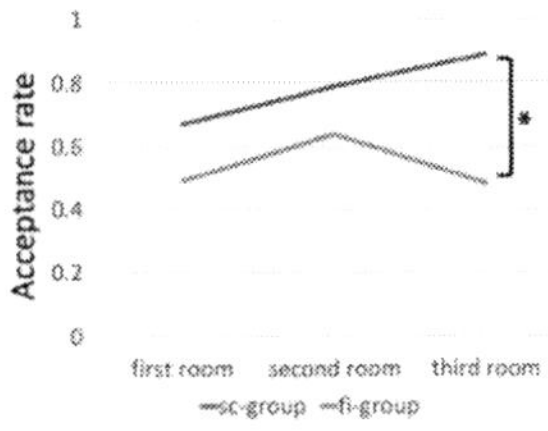

Figure 6. Acceptance rates of convergent metacognitive suggestions.

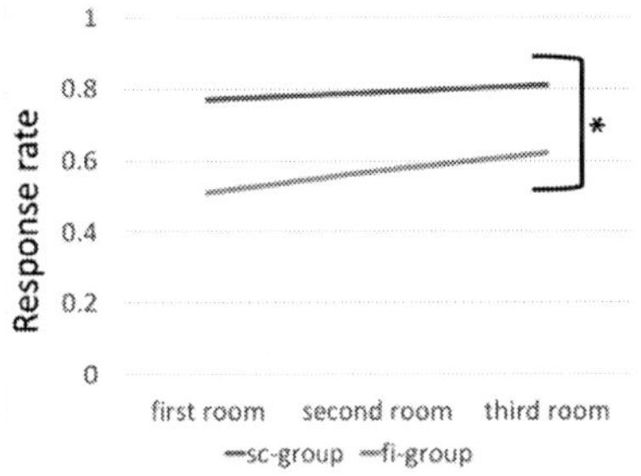

Figure 7. Rates of suggestions with LF/HF responses during the 30 s after the suggestion.

IV. DISCUSSION

We hypothesized that participants would be likely to accept the advice based on the estimation of the inner state of the human even when the agent provided advice. In this research, we focused on metacognitive suggestion in an insight problem-solving task, which is one of the examples of useful advice that the agent can provide. We investigated the effects of metacognitive suggestion that controlled the timing of presentation based on human inner state. We implemented an agent that estimated two kinds of user states: a thinking mode (depth-first or breadth-first) and a state of the stalemate (impasse or progressive). The agent categorized the participant's overall state as one of four combinations: depth-first/progressive, breadth-first/progressive, depth-first/impasse, and breadth-first/impasse. The agent provided a metacognitive suggestion with the goal of getting humans out of the impasse state.

We conducted an experiment using two suggestion agents. One was a state-considering agent that estimated the user's state before providing a metacognitive suggestion. Another was a fixed-interval agent that provided a metacognitive suggestion in three-minute intervals. Based on results from the analysis of operation history log, the acceptance rate of suggestions in the sc-group was significantly higher than that in the fi-group. In other words, the attitude to the metacognitive suggestions given by the agent was different between the participants in fi-group and those in the sc-group. The participants in the sc-group believed that the content of the suggestions given by the agent should be considered. On the other hand, participants in the fi-group typically thought that the agent's suggestions presented general knowledge, and might accept useful ones regardless of the task status. This is also suggested from the fact that participants' physiological index (LF/HF) were often responsive to the agent's metacognitive suggestions in

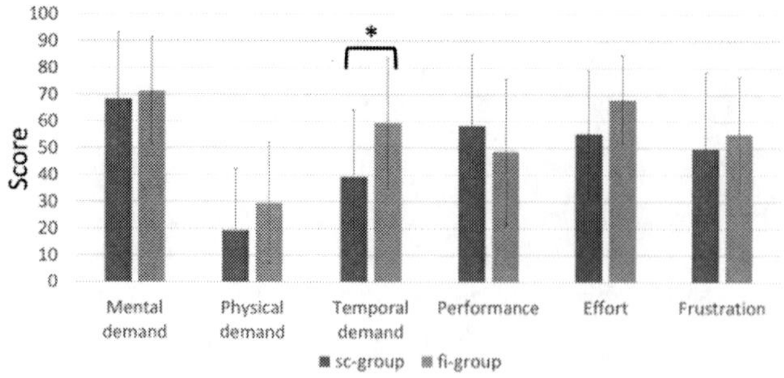

Figure 8. Results of mental workload measurements.

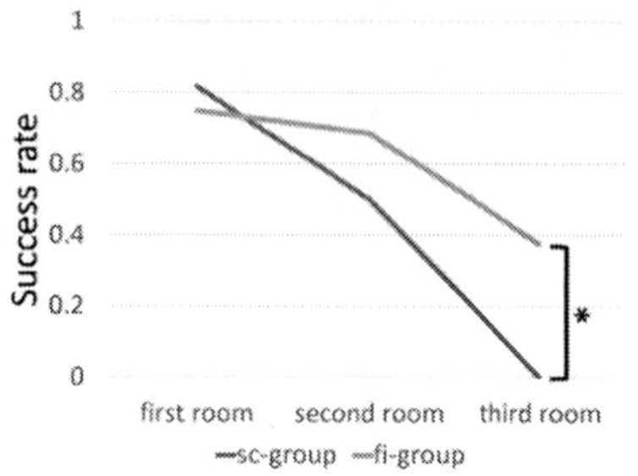

Figure 9. Results of success rates of escaping from the rooms.

the sc-group. The results of the mental workload suggest that participants in the fi-group might interpret the agent's suggestions as a kind of facilitation of the task execution rather than human assistance.

There was an unexpected result. In the third room, the participants in the sc-group could not escape from the room. The total success rate in the sc-group was lower than that in the fi-group. We expect that one of the reasons is that the search space of the third room was larger other rooms. The third room included many game objects and the process of problem solving was complex. Therefore, even when the impasse estimation system judged that the breadth-first search state of the participants continued too long, the participant often needed to continue to explore the search space to find a way to escape the room. Although it is not necessary to understand the contents of the task in order to make appropriate interventions by metacognitive suggestion, it is necessary to decide whether to encourage divergence or convergence, taking into consideration the size of the search space of the task. In this study, we focused on the human inner state to decide the content and timing of the suggestion. However, in a general situation, it is necessary to consider a method of giving advice by adding relatively abstract information that expresses the feature of the task, such as the size of the search space. In addition, because the agent's suggestions were appropriate to some extent until the second room, participants in the sc-group might fall into an assistant dilemma [13].

V. Conclusions

The aim of this study is to investigate whether the acceptance of metacognitive suggestions provided by an agent could be encouraged by controlling the content and timing of suggestion presentation according to the state of the participant. We implemented an agent that estimated two kinds of user states: a thinking mode (depth-first or breadth-first) and a state of the stalemate (impasse or progressive). From the experiment using the two types of suggestion agents, we could suggest that participants was likely to accept the metacognitive suggestions provided by the agent when the suggestions were given based on the inner state estimation of the participant. However, the task performance decreased. We speculate that this might have caused the agent to give inappropriate content suggestions, and then causing confusion to the participant because the participant was accepting the agent's suggestions. In a future work, we will improve the system to incorporate the task feature estimation (this is not content recognition).

Acknowledgment

This research is supported by Grant-in-Aid for Young Scientists (B) (KAKENHI No. 16K21113), and Grant-in-Aid for Scientific Research on Innovative Areas (KAKENHI No. 26118002) from the Ministry of Education, Culture, Sports, Science and Technology of Japan.

References

[1] T. Okada and H. A. Simon, "Collaborative discovery in a scientific domain," Cognitive science, vol. 21, no. 2, 1997, pp. 109–146.

[2] K. Ueda and K. Niwa, "Cognitive analysis of collaborative knowledge creation in r&d teams," in Innovation in Technology Management. The Key to Global Leadership. PICMET'97. IEEE, 1997, pp. 524–527.

[3] R. G. Hass, "Effects of source characteristics on cognitive responses in persuasion," Cognitive responses in persuasion, 1981, pp. 141–172.

[4] J. Patrick, A. Ahmed, V. Smy, H. Seeby, and K. Sambrooks, "A cognitive procedure for representation change in verbal insight problems." Journal of Experimental Psychology: Learning, Memory, and Cognition, vol. 41, no. 3, 2015, p. 746.

[5] Y. Hayashi, "Social facilitation effects by pedagogical conversational agent: Lexical network analysis in an online explanation task." International Educational Data Mining Society, 2015.

[6] S. T. Iqbal and B. P. Bailey, "Leveraging characteristics of task structure to predict the cost of interruption," in Proceedings of the SIGCHI conference on Human Factors in computing systems. ACM, 2006, pp. 741–750.

[7] K. Isbister, H. Nakanishi, T. Ishida, and C. Nass, "Helper agent: Designing an assistant for human-human interaction in a virtual meeting space," in Proceedings of the SIGCHI conference on Human Factors in Computing Systems. ACM, 2000, pp. 57–64.

[8] Y. Ohmoto, H. Sonobe, and T. Nishida, "Controlling the timing of metacognitive suggestions using impasse estimation," in Proceedings of the 6th International Conference on Human-Agent Interaction. ACM, 2018, pp. 353–355.

[9] S. Ohlsson, "Information-processing explanations of insight and related phenomena," Advances in the psychology of thinking, vol. 1, 1992, pp. 1–44.

[10] H. Shirouzu, N. Miyake, and H. Masukawa, "Cognitively active externalization for situated reflection," Cognitive science, vol. 26, no. 4, 2002, pp. 469–501.

[11] Y. Hayashi, "Togetherness: Multiple pedagogical conversational agents as companions in collaborative learning," in International Conference on Intelligent Tutoring Systems. Springer, 2014, pp. 114–123.

[12] Y. Ohmoto, T. Matsuda, and T. Nishida, "Experimentally analyzing relationships between larner's status in the skill acquisition process and physiological indices." International Journal on Advances in Life Sciences, vol. 9, no. 3 and 4, 2017, pp. 127–136.

[13] K. R. Koedinger and V. Aleven, "Exploring the assistance dilemma in experiments with cognitive tutors," Educational Psychology Review, vol. 19, no. 3, 2007, pp. 239–264.

EEG Application for Human-Centered Experiments in Remote Ship Operations

Raheleh Kari
Department of Ocean Operations and Civil Engineering
Norwegian University of Science and Technology (NTNU)
Ålesund, Norway
e-mail: raheleh.kari@ntnu.no

Martin Steinert
Department of Mechanical and Industrial Engineering
Norwegian University of Science and Technology (NTNU)
Trondheim, Norway
e-mail: martin.steinert@ntnu.no

Henrique M.Gaspar
Department of Ocean Operations and Civil Engineering
Norwegian University of Science and Technology (NTNU)
Ålesund, Norway
e-mail: henrique.gaspar@ntnu.no

Abstract— **Electroencephalography (EEG) is an exceptional technique in order to study human behavior via neurocognitive processes. EEG can be utilized in different domains to study the neurocognitive processes behind human-machine interactions. Furthermore, remote ship operations are an innovative approach in the maritime industry to improve the efficiency of operations. However, in the Shore Control Center (SCC), human-machine interactions are challenging and highly important in order to implement successful remote ship operations. In this respect, human operators are key components of the SCCs; performance and efficiency of their interactions directly affect remote monitoring and remote controls. Hence, the present study introduces the EEG EPOC Flex in order to objectively investigate human interactions. The current paper focuses on operator's stress in different levels of workload within the SCC. In effect, EEG will facilitate investigating human factor issues that may affect operators' interactions in the SCC. The result of this study can help SCC designers to design an efficient environment for SCC operators.**

Keywords-Remote operations; Ship; EEG; Human experiments

I. INTRODUCTION

Today, studying the underlying mechanisms of human behavior is the center of interest for many researchers in different fields, especially for automated environments and remote-controlled vessels. Thus, different types of Human Factor Engineering (HFE), Human-Computer Interaction (HCI), and affective engineering experiments have been performed [1], [2]. While the concept of remote-controlled vessels and remote ship operations are in their early stage, study the human interactions will significantly improve the functionality and value of remote ship operations. Although the autonomous vessel might be in the form of an unmanned vessel, human operators still play an important role in the unmanned systems [3] within the SCC. In this respect, operators' stress due to information overload is one of the important identified human factor issues; because one SCC operator has to deal with several vessels at the same time [3], [4]. According to the communication requirements for an unmanned vessel, remote control and remote monitoring are two important control modes within the SCC [5] where humans play a critical role to monitor and control the vessels from distant locations.

The concept of SCC relies on a manned shore-based control center which is responsible for the ship operation. In this scenario, difficult or critical remote ship operations will be conducted by the SCC [5]. However, in order to decrease the workload on the SCC operators, assure the safety, or in the case of the limited or loss of the communication between the vessel and SCC, the vessel must be able to perform the normal operation without human supervision. There are several fails to safe strategies that are programmed and updated by the SCC and the vessel can enter one of the possible strategies when SCC cannot respond [5].

According to the concept of remote ship operation and SCC, the crew will be relocated from on board the vessel to shore-based control centers, and they will perform different tasks based on the organization and control modes of the SCC. This relocation does not mean the elimination of the human factors or solving all human errors; by contrast, more questions regarding human factors and human interactions will be raised, because SCC operators have to take the full control over the vessel at any time [4]. In addition, the SCC operator is defined as an Officer of Watch (OOW), who is responsible for monitoring the vessel at any time and interevent if needed [6]. This is highlighted in the remote monitoring mode of SCC as well [5].

Thus, human operators are key components of the SCC and their performance and efficiency directly affect remote ship operations. This implies that study the human-human and human-machine interactions are highly important in order to implement remote ship operations. In this respect, Electroencephalography (EEG) is a special tool that can be utilized for quantifying human interactions and study the neurocognitive processes underlying human behavior [7]. Although there are various studies that investigate the behavioral interactions in a qualitative manner, using physiological sensors especially EEG leads to quantify human interactions. EEG records brain waves and electrical activities by using electrodes on the scalp. Electrical activities of the brain indicate that how is the communication between different neurons in the brain network through electrical impulses [8].

The current study presents the application of EEG EPOC Flex in order to objectively investigate human factor issues affecting SCC operators. According to [9], there are different types of issues that affect human interactions during the remote ship operations especially within the SCCs. For example, a high level of mental workload and a lack of direct sensory. On the other hand, the main hypothesis of the successful implementation of the Human-Machine Interface (HMI) within the SCC introduces an SCC operator as a person who has to be able to monitor and control up to six vessels at the same time [10]. In this respect, the current paper focuses on using EEG EPOC Flex for measuring the stress of SCC operators during the remote monitoring and remote ship operations in different levels of workload. The paper presents stress as a dependent variable and workload as an independent variable, which can be manipulated during an EEG experiment.

Developing this type of human-centered experiment leads to the study of human interactions in a quantitative manner, which can boost the acceptability of results. But, the development process of a human-centered experiment requires fundamental knowledge in various domains, such as psychology, physiology, engineering design, electro-physics, and neuroscience [2]. This increases the complexity of the development process, especially when the experiment is to be designed from scratch and there is no clear experiment precedes for scenarios [2]. Thus, the current paper reviews some aspects of knowledge domains in order to build an EEG human-centered experiment within the SCC. In addition, an example case comprising a high workload and low workload scenarios is presented. The example case is developed based on the wayfaring model; this model supports the early concept creation stage of designing experiments including multi-disciplines and high levels of complexity [2].

The rest of the paper is organized as follows: the definition of remote ship operations and SCC in section II. Electroencephalography (EEG) and its characteristics appear in section III; this section reviews how stress can be investigated by using brain frequencies. Wayfaring model and example case is presented in section IV. Section V discusses the objective and EEG as an explicit tool to design a human-centered experiment regarding the stress of SCC operators. Finally, section VI presents the conclusion and future works.

II. REMOT SHIP OPERATIONS AND SCC

Today, technology is generally available to operate a vessel without on board crew, however, several developments and validations are required [5]. According to the definition of remote operation, a machine is under the continuing and direct human control from a distant location [11]. This implies that the concept of unmanned vessels will rely on an SCC, which is responsible for the operation of the vessel based on the different types of vessel and SCC control modes [5].

According to the baseline case from Maritime Unmanned Navigation through Intelligence in Networks (MUNIN), the SCC contains all on land functions, remote bridge, and engine control modules that can be utilized to take the direct control over the vessel in specific cases [12]. Furthermore, the initial voyage planning will be conducted in SCC, and all the voice communication to the vessel will be relayed to the SCC in order to handle by the SCC operators [12].

There are five main ship control modes for an onboard Autonomous Ship Controller (ASC) comprising autonomous execution, autonomous control, direct remote control, indirect remote control, and fail to safe mode [5]. According to different modes of ship control, it can be concluded that SCC will operate in various modes as well. Remote monitoring, remote operation, status investigation, ASC update, and intervention are the main SCC control modes [5]. Generally, the SCC will be in the remote monitoring mode, when all the ship status indicators are normal; in this case, no action will be taken from SCC. The ASC sends a set of ship status flags to SCC at short intervals (for example every 5 seconds). A set of ship status indicators are presented by [5] based on the hierarchical ship function decomposition [13] and the ship functional model [14]. Table I indicates ship status indicators which are critical parameters for the SCC operators in different SCC control modes.

TABLE I. SHIP INDICATORS IN SCC [5]

Indicator	Detailed description elements
Location	Position, heading, speed, distance from planned position as well as position quality flag.
Weather	Wind speed/ direction, wave and swell hight/ length/ direction
Visibility	Visibility, radar range and culture. COLREG status of ship
Collision	Vectors to targets, status/heading/ speed of targets
Grounding	Depth measurement
Communication	Critical communication directly to ship, for example on VHF, GMDSS.
Stability	Trim, heel, draft, water tight integrity, void space, water ingress
Environment	Environmental performance and emission to air and sea
Economy	Fuel use and potential for late arrival/ off hire etc
Hull propulsion	Hull and equipment status, anchor, towing, ladders
Propulsion	Engine, auxiliaries, piping, fuel
Electric	Electric power systems, switch boards, emergency power
Safety	Fire, evacuation, extinguishing, escape
Cargo	Cargo status

If one of the above indicators shows some abnormality, the SCC will enter to ASC update and investigation modes [5]. Remote ship operation mode is utilized to control the vessel manually when ASC cannot solve the problem in a certain situation. SCC will enter intervention mode when deeper interactions with onboard systems are necessary [5], [12].

III. ELECTROENCEPHALOGRAPHY (EEG)

Electroencephalography in short EEG means recording the electric activity of the brain. Most cognitive processes happen within tens to hundreds of milliseconds, this is much faster than the blink of an eye. On the other hand, events, which trigger cognitive processes happen in time sequences that span hundreds of million seconds to a few seconds [8]. EEG similar to a high-speed camera has a high time resolution; hence, it can capture the physiological changes of underlying cognitive processes better than other brain imaging techniques including Positron Emission Tomography (PET) scanners and Magnetic Resonance Imaging (MRI) [8].

The brain is continually active and generates electrical activities that are significantly less than a 9v battery. EEG sensors can pick up weak signals from the surface of the scalp [8]. The international brain research has been gaining important findings regarding EEG and established theories which are well-accepted on how the EEG signals relate to attentional, affective or cognitive processing [8]. The current paper utilized the Emotiv EEG EPOC Flex containing 32 channels. Each sensor records five main brain band powers consisting Theta (4-8Hz), Alpha (8-12Hz), Low beta (12-16Hz), High beta (16-25Hz), and Gamma (25-45Hz). EEG researchers record the brain's electrical activities via sensors placed at the scalp surface; this applied first to humans by a German neurologist Hans Berger in the 1920s [8]. As the electrical signals are very small, the data is digitized and sent to an amplifier. Then, the recorded data can be displayed as a time series of voltage values [8].

American Encephalographic society (1994) has been provided the most common system in order to define and name the electrode locations/positions on the scalp. This system is called the 10-20 system such that electrodes are placed at 10% and 20% points along lines of longitude and latitude. In the 10-20 system electrode names start with one or two letters showing the general brain region or lobes where the electrode is placed [8].

Electrode's name end with a letter or a number showing the distance to the midline. Even numbers in the right hemisphere, and odd numbers in the left hemisphere. Larger numbers refer to the greater distances from the midline [8]. Electrodes placed at the midline are labeled with "z", which indicates zero distance from the midline. For instance, T7 is placed over the left temporal regions, Fp8 is placed over the right front-polar regions, and Cz is located over the midline central brain region [8].

A. EEG frequencies

Brain oscillations with the 4-8 Hz frequency range are identified as theta band [15], [16]. Various studies highlight that frontal theta activities are correlated with the difficulty of mental operations; for instance, during learning, information takes up, focused attention or memory recall. In this respect, theta frequencies are more prominent when the task difficulty increase. Hence, theta is usually associated with the processes underlying working memory and mental workload [8], [17]–[19]. Workload refers to any cognitive process involving executive processes including problem-solving, working memory and analytical reasoning. Workload which is associated with the theta band power, increases in the case of a higher level of task demand and working memory load [8]. Theta frequency can be recorded from all over the cortex, this indicates that it is generated via a wide network involving central, parental, prefrontal and temporal cortices [8]. On the other hand, there is an improvement of theta waves in the case of mental stress. This implies a good correlation between EEG signals and mental stress [20], [21]. Table II indicates a summary of studies on theta and other EEG bands focusing on stress and workload.

Alpha band as a rhythmic oscillatory activity with the frequency range between 8–12 Hz was discovered by Hans Berger in 1929 [15] cited by [8]. The alpha band is generated in posterior cortical areas, comprising parietal, occipital and posterior temporal brain areas [8], however, it is more active in frontal and occipital regions of the brain [22]. Figure 1 indicates four main brain lobes and their responsibilities. In normal adults, alpha waves can be seen spontaneously during wakefulness and relaxed state [18]. The alpha band is correlated with sensory, memory, and motor functions. During the mental and physical relaxation with closed eyes, there is an increase in alpha band power. By contrast, alpha power is suppressed during mental and body activities with open eyes [8]. In other words, alpha suppression indicates that the brain is preparing to pick up information from different senses, focusing on what really matters in that specific moment, and coordinating attentional sources [8]. This implies that alpha suppression is a valid presentation of engagement and mental activities, for instance during focused attention to different types of the stimulus [23] cited by [8].

Importantly, in the case of no stress, when the brain is performing no activity alpha waves are dominant. In stressful conditions, the alpha power declines to indicate the change in the response under stress [21]. The power suppression of alpha band during the stressful situation is also highlighted by [24]. It is important to mention that in a study regarding functional roles of 10 Hz alpha band power, Magnetoencephalography (MEG) signals similar to EEG signals indicate that alpha oscillations are associated with semantic, spatial and social attention [25].

Oscillations within the range of 12-25 Hz are generally referred to as beta band activities [8], [16], [18]. Beta frequency is generated in both frontal and posterior regions of the brain. Higher beta power is generally correlated to active concentration and busy or anxious thinking [8], [32].

TABLE II. SUMMARY OF EEG STUDIES ON STRESS AND WORKLOAD

Authors	EEG band	EEG sensors	Results
[26]	Delta, Theta, Alpha, Beta, Gamma	F_3,FC_5,T_7,P_7,O_1,O_2, P_8,T_8,FC_6,F_4,F_8,AF_3, F_7, AF_4	The results indicate the feasibility of EEG in order to detect stress. The highest accuracy was obtained in the alpha band power.
[27]	Theta (fmθ) Beta	71 EEG electrodes	Theta (fmθ) increased consistently with memory load.
[28]	Beta	FC_5, FC_6, O_1, O_2,	Participants in the stress group had the highest level of Beta activity.
[29]	Theta	The locations of 48 electrodes (out of 171) are shown in the study	In more complex mazes there are more theta oscillations. Theta oscillations are more frequent through recall trials compare with learning trials.
[20]	Theta, Beta, Delta, Alpha,	FP_1,FP_2, F_3, F_4,C_3,C_4,P_3, P_4,O_1,O_2, F_7,F_8,T_3,T_4, T_5,T_6,Fz,Cz, Pz, Oz	During the most complex task, delta power increased. Delta only increases when attention to internal processing is required. During the performance of a task, alpha activity was suppressed, but low theta increased.
[23]	Alpha	C_4-A_2,C_3 A_1,O_2- A_2,O_1-A_1	During the visual stimulation, a decreased in Rhythmic Activity within the Alpha Band (RAAB) happened in occipital regions. During the auditory stimulation, power declined in the central regions.
[30]	Alpha, Beta, Theta	Fp1,AF3, F3,F7,FC5, FC1,C3,T7, CP5, CP1,P3,P7, PO3,O1, Oz,Pz,Fp2, AF4,Fz,F4, F8,FC6,FC 2,Cz,C4,T8, Cp6,Cp2,P 4,P8,PO4, O2	In the field of spectral centroids, 30 out of 32 electrodes indicated increased centroid frequency during stress (without significant effect). It is argued that negative valence stimuli lead to an increase in the right frontal activities. Besides, left frontal activity indicates the EEG response to positive valence stimuli.
[31]	Theta, Alpha, Beta	Fp1,Fp2, FpZ(prefrontal sites)	Stressful subjects showed a higher level of EEG complexity. The group with chronic stress had higher left prefrontal power. There was a significant difference in relative power of alpha between the normal group and the stress group.
[19]	Theta, Alpha	19 electrodes based on the 10-20 system (including O_2, P_4, F_7)	Upper Alpha is dominant during retention (increasing the memory load). Theta and upper Alpha play an important role during the retrieval.

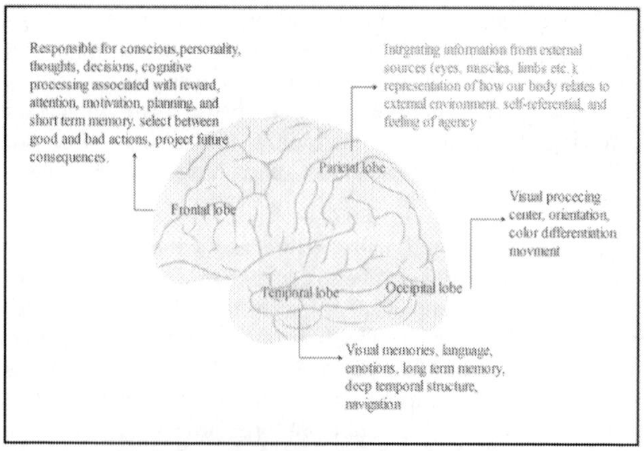

Figure 1. Brain lobes and their responsibilities

Another study indicates that there is a high beta activity at the FC_5 electrode during the stress situation [28]. This means that there is a meaningful correlation between beta frequencies and levels of stress.

Oscillations above 25 Hz are referred to gamma frequencies which usually found during conscious perception. Due to high contamination by muscle artifacts and small amplitude [18]. Gamma frequencies are black holes of the EEG studies. Still, it is unclear that which parts of the brain generate gamma frequencies and what these frequencies reflect. Some studies argue that the gamma band is a by-product of other neural processes, hence, they do not reflect any cognitive processing at all [8]. Notably, high gamma activity in the temporal region is associated with memory processes [18]. Furthermore, some studies report that gamma activity is connected to attention, working memory and long-term memory processes [18], [33].

IV. HUMAN-CENTERED EXPERIMENT IN THE EXAMPLE CASE

This part presents an example of a human-centered experiment to study stress and workload within the SCC. The wayfaring approach is applied to the development of the experiment. The current example comprises two scenarios with different levels of workload. The research hypothesis is defined as follows:

H_1: There is a significant difference in participant's stress between high workload and low workload scenarios.

A. Wayfaring approch

Wayfaring model [34], [35] can be utilized as an exploration journey instead of a planning-based approach to discover innovative ideas. The wayfaring model describes that "an optimum new solution to a problem cannot be preconceived as we do not have empirical evidence for the outcome of something that has not previously been done" [2]. In this respect, the model provides a methodology for a practical exploration of the problem and solution. According to [2], this methodology comprises four steps as follows:

1. Probing ideas: the exploration of opportunities, sometimes this phase includes a low-resolution prototype in order to fail early and to abductive learning.
2. Merging multidisciplinary: comprising all knowledge domains from the beginning to discover interdependencies and develop interrelated knowledge
3. Speed: make a plan based on the short iteration timeframe to maximize the iteration numbers
4. Agility: opportunistically select the next step and let the development process shape the outcome. This phase can be followed in order to achieve serendipity findings and innovation outcomes

B. *Scenario 1*

This experiment uses EEG EPOC Flex to record the brain activity of a human operator while monitoring the vessel's operation in the simulation environment. The task takes 10 minutes due to limitations for using EEG EPOC Flex. The instructor room of the navigation simulators will be considered as an SCC and a ship bridge as a remote-controlled vessel. As we can see in Figure 2, the instructor room is assumed as an SCC where one operator can monitor up to six vessels (SCC remote monitoring mode). In the current scenario, an SCC operator is defined as an OOW. As mentioned before, this person is responsible for monitoring the ship and intervening if required [6]; this means sending high-level commands in specific situations. The ship bridge will be operated by simulator assistants based on the predefined route. In this scenario, SCC operators will fill in a self-report stress survey after they finish the experiment. In addition, the paper-based NASA TLX forms will be used in order to assess the workload objectively.

In scenario number 1, SCC operator will monitor one remotely operated vessel from the instructor room (based on the monitor mode of SCC), and there is a ship-ship communication between the bridge operators and SCC operators. SCC operator is responsible to send high-level commands if needed in the case of accidents, crossing other vessels or harsh weather. In addition, SCC operator has to monitor the radar data, heading, propeller revolution, rate of turn, rudder and speed of each vessel to ensure that all the vessels are in an appropriate status. Independent variable will be manipulated as follows:

- Area: Kristiansund to Trondheim (low difficulty)
- One container ship as the main vessel
- 5+ targets (traffic)
- Good visibility in daylight
- Moderated wind, calm sea-state
- No accidental situation

Figure 2. Simulator instructor room

C. *Scenario 2*

Actors, environment, and components of scenario number 2 are the same as scenario number 1. The independent variable will be manipulated as follows:

- Area: Vatlestraumen (moderate difficulty)
- 5 container ships as the main vessels
- 15+ targets (traffic)
- Bad visibility, night time
- Strong wind, wavy sea
- Two accidental situation

It is notable that in this experiment physiological baseline will be performed before scenario 1.

V. DISCUSSION

Remote monitoring and remote controls have been merged in different types of control modes of unmanned vessels and SCCs. In this respect, SCC appears as a black box comprising human, hardware, software and different types of human-machine interfaces. It is clear that by relocating crew from on board the vessel to SCC, we will face more human factor challenges such as stress and information overloading. Hence, study human interactions within the SCC leads to gain more knowledge about different aspects of remote ship operations and decoding the black box of the SCC. The current paper presented EEG as a remarkable method in order to investigate the stress of SCC operators during the different levels of workload.

Moreover, the paper presented a very simple example case based on the wayfaring approach. The wayfaring model was an appropriate model in order to develop the experiment due to a lack of clear procedures and a high level of complexity. EEG EPOC Flex was defined as a tool to record the participant's brain signals during the experiment. According to the various studies regarding the EEG signals and stress, we suggest theta and alpha frequencies as two important EEG frequencies which can be considered to study stress. During the stressful situation, the power of alpha frequencies falls down, however, the power of theta waves increases. In addition, EEG electrodes in the frontal region of the brain including Fp1, Fp2, FpZ (prefrontal site), FC_5 and FC_6 can be considered as effective electrodes during the stress evaluation. However, a comprehensive analysis of stress requires considering more EEG electrodes in addition to different band powers.

Developing a human-centered experiment is a challenging task, as fundamental knowledge in the different domains is required. This implies one of the main challenges in developing human-centered experiments, especially in the field of EEG. Using a simulator environment rather than a real SCC is another issue, which can affect the results of the experiment. Furthermore, the availability of participants is another limitation during the human-centered experiments. It is important to mention that in the current high workload scenario, all the independent factors (number of targets, weather, route, etc.) were manipulated aiming to make a significant change in the level of workload; thus, the effect of each factor is not clear. Future experiments can provide more details regarding the independent variable.

VI. CONCLUSION

In this paper, we studied general aspects of remote ship operations, SCC and EEG in order to a develop human-centered experiment. EEG is presented as an exceptional tool to study stress in different levels of workload within the SCC. Moreover, an example case including high and low workload scenarios was defined by applying the wayfaring approach. The current example case focused on the monitoring mode of the SCC. In this respect, future studies can investigate stress and workload in other SCC control modes. Moreover, future studies can investigate different analysis and pre-processing techniques in order to extract the required data of stress and workload from raw EEG data.

REFERENCES

[1] S. Balters and M. Steinert, "Capturing emotion reactivity through physiology measurement as a foundation for affective engineering in engineering design science and engineering practices," *Journal of Intelligent Manufacturing*, vol. 28, no. 7, pp. 1585–1607, 2017.

[2] K. K. Leikanger, S. Balters, and M. Steinert, "Introducing the wayfaring approach for the development of human experiments in interaction design and engineering design science," in *DS 84: Proceedings of the DESIGN 2016 14th International Design Conference*, 2016, pp. 1751–1762.

[3] T. Porathe, J. Prison, and M. Yemao, "Situation awareness in remote control centres for unmanned ships," p. 9, 2014.

[4] Y. Man, M. Lundh, and T. Porathe, "Seeking Harmony in Shore-Based Unmanned Ship Handling: From the Perspective of Human Factors, What Is the Difference We Need to Focus on from Being Onboard to Onshore?," in *Human Factors in Transportation*, CRC Press, 2016, pp. 81–90.

[5] Ø. J. Rødseth, B. Kvamstad, T. Porathe, and H. Burmeister, "Communication architecture for an unmanned merchant ship," in *2013 MTS/IEEE OCEANS - Bergen*, 2013, pp. 1–9.

[6] Ø. J. Rødseth and H. avard Nordahl, "Definitions for autonomous merchant ships," in *Norwegian Forum for Unmanned Ships*, 2017.

[7] M. X. Cohen, "It's about Time," *Front. Hum. Neurosci.*, vol. 5, 2011.

[8] iMotions, "EEG (Electroencephalography): The Complete Pocket Guide," *iMotions*, 07-Jun-2016. [Online]. Available: https://imotions.com/blog/eeg/. [Accessed: 27-May-2019].

[9] R. Kari, H. M. Gaspar, A. H. Gausdal, and M. Morshedi, "Human Interactions Framework for Remote Ship Operations," in *2018 26th Mediterranean Conference on Control and Automation (MED)*, 2018, pp. 1–9.

[10] S. N. MacKinnon, Y. Man, and M. Baldauf, "D8.8: Final Report: Shore Control Centre." Maritime Unmanned Navigation through Intelligence in Networks, 2015.

[11] T. B. Sheridan, *Telerobotics, automation, and human supervisory control*. MIT press, 1992.

[12] Ø. J. Rødseth, Å. Tjora, and P. Baltzersen, "D4.5: Architecture specification." MUNIN, 2014.

[13] Ø. J. Rødseth *et al.*, "TCI and status indicator specification, Flagship deliverable D-D1", 2008. [Online]. Available: http://www.mits-forum.org/resources/D1-1-TCIs-v12.pdf [Accessed: 29-May-2019].

[14] V. Vindøy, "A functionally oriented vessel data model used as basis for classification," in *7th International Conference on Computer and IT Applications in the Maritime Industries, COMPIT*, 2008, vol. 8.

[15] E. Niedermeyer and F. H. L. da Silva, *Electroencephalography: Basic Principles, Clinical Applications, and Related Fields*. Lippincott Williams & Wilkins, 2005.

[16] D. L. Schomer and F. L. da Silva, *Niedermeyer's Electroencephalography: Basic Principles, Clinical Applications, and Related Fields*. Lippincott Williams & Wilkins, 2012.

[17] W. Klimesch, "Memory processes, brain oscillations and EEG synchronization," *International journal of psychophysiology*, vol. 24, no. 1–2, pp. 61–100, 1996.

[18] A. S. Malik and H. U. Amin, "Chapter 1 - Designing an EEG Experiment," in *Designing EEG Experiments for Studying the Brain*, A. S. Malik and H. U. Amin, Eds. Academic Press, 2017, pp. 1–30.

[19] B. Schack, W. Klimesch, and P. Sauseng, "Phase synchronization between theta and upper alpha oscillations in a working memory task," *International Journal of Psychophysiology*, vol. 57, no. 2, pp. 105–114, Aug. 2005.

[20] T. Harmony *et al.*, "EEG delta activity: an indicator of attention to internal processing during performance of mental tasks," *International journal of psychophysiology*, vol. 24, no. 1–2, pp. 161–171, 1996.

[21] A. R. Subhani, L. Xia, and A. S. Malik, "EEG signals to measure mental stress," in *2nd International Conference on Behavioral, Cognitive and Psychological Sciences, Maldives*, 2011, vol. 2012, pp. 84–88.

[22] S. Tong and N. V. Thakor, *Quantitative EEG analysis methods and clinical applications*. Artech House, 2009.

[23] G. Pfurtscheller and A. Aranibar, "Event-related cortical desynchronization detected by power measurements of scalp EEG," *Electroencephalography and clinical neurophysiology*, vol. 42, no. 6, pp. 817–826, 1977.

[24] C. Setz, B. Arnrich, J. Schumm, R. La Marca, G. Tröster, and U. Ehlert, "Discriminating stress from cognitive load using a wearable EDA device," *IEEE Transactions on information technology in biomedicine*, vol. 14, no. 2, pp. 410–417, 2009.

[25] K. D. Rana and L. M. Vaina, "Functional Roles of 10 Hz Alpha-Band Power Modulating Engagement and Disengagement of Cortical Networks in a Complex Visual Motion Task," *PLoS ONE*, vol. 9, no. 10, p. e107715, Oct. 2014.

[26] V. Vanitha and P. Krishnan, "Real time stress detection system based on EEG signals," 2016.

[27] J. Onton, A. Delorme, and S. Makeig, "Frontal midline EEG dynamics during working memory," *Neuroimage*, vol. 27, no. 2, pp. 341–356, 2005.

[28] S.-H. Seo and J.-T. Lee, "Stress and EEG," in *Convergence and hybrid information technologies*, IntechOpen, 2010.

[29] M. J. Kahana, R. Sekuler, J. B. Caplan, M. Kirschen, and J. R. Madsen, "Human theta oscillations exhibit task dependence during virtual maze navigation," *Nature*, vol. 399, no. 6738, p. 781, 1999.

[30] G. Giannakakis, D. Grigoriadis, and M. Tsiknakis, "Detection of stress/anxiety state from EEG features during video watching," in *2015 37th Annual International Conference of the IEEE Engineering in Medicine and Biology Society (EMBC)*, 2015, pp. 6034–6037.

[31] H. Peng *et al.*, "A method of identifying chronic stress by EEG," *Personal and ubiquitous computing*, vol. 17, no. 7, pp. 1341–1347, 2013.

[32] Y. Zhang, Y. Chen, S. L. Bressler, and M. Ding, "Response preparation and inhibition: the role of the cortical sensorimotor beta rhythm," *Neuroscience*, vol. 156, no. 1, pp. 238–246, 2008.

[33] C. S. Herrmann and T. Demiralp, "Human EEG gamma oscillations in neuropsychiatric disorders," *Clinical Neurophysiology*, vol. 116, no. 12, pp. 2719–2733, Dec. 2005.

[34] T. Ingold, *Lines : A Brief History*. Routledge, 2016.

[35] M. Steinert and L. J. Leifer, "'Finding One's Way': Re-Discovering a Hunter-Gatherer Model based on Wayfaring," *International Journal of Engineering Education*, vol. 28, no. 2, p. 251, 2012.

Designing Effective Advergame for Purchase Intention: A View of Game Design and Psychological State

Wei-Hung Hsiao
Department of Distribution Management
National Taichung University of Science & Tech.
Taichung, Taiwan
e-mail: weihung@nutc.edu.tw

Ing-Long Wu
Department of Information Management
National Chung Cheng University
Chia-Yi, Taiwan
e-mail: ilwu@mis.ccu.edu.tw

Shwu-Ming Wu
Department of Human Resource Management,
National Kaohsiung University of Science & Tech.
Kaohsiung, Taiwan
e-mail: mingwu@cc.kuas.edu.tw

***Abstract*-Advergame has increasingly been used for a new marketing campaign. This study defines a new research model with a relationship structure of advergame design and individual psychological state as antecedents for advergame attitude, brand attitude, and purchase intention, based on the hierarchy of effects theory. An empirical study for online survey collects 522 consumers who have recent experience of advergame. The results found that game-brand fit, perceived interactivity, and involvement are the key drivers of advergame attitude. Further, advergame attitude and brand attitude were found to be two important mediators of purchase intention. Implications for marketers and scholars are further discussed.**

Keywords-Advergame; advergame design; advergame attitude; brand attitude; purchase intention.

I. INTRODUCTION

Advergame is a particular form of branded entertainment that the brands, along with related messages, logos, and trademarked characters, are included in a game, and such promotional tools are usually accessed via the parent companies or brand websites [27]. Advergame has increasingly become one of the most promising areas of Internet advertising in recent years [33]. Many companies have tried advergame as part of their marketing strategies, such as Coca-Cola, Toyota, Michelin, and McDonald.

Most studies have focused on the issues related to advertising effectiveness, such as brand/product recognition or attitude [11] and advergame design [6][14]. However, an important role of advergame in marketing is persuasive purchase behavior, which remains relatively under-explored and this is one gap in the literature that the current work aims to address. Moreover, researchers noted that advertising is generally seen as a means of communicating persuasively with consumers [26] and that if it is successful, it will ultimately result in consumers purchasing the focal branded items. It is imperative to examine the issue of purchase intention in relation to advergame to assess marketing effectiveness.

Lavidge and Steiner [19] proposed the hierarchy of effects theory for predicting advertising effectiveness, and its basic premise is that advertising moves people up a series of steps toward an actual purchase behavior. These steps can be further classified as three functions related to a classic psychological model, that is, cognitive, affective, and conative [1]. Many studies on Internet advertising have proposed a similar stage-based structure based on this theory, that is, belief, attitude, and behavior, to examine different forms of advertising effectiveness, such as brand loyalty and sale promotion [2]. Many studies have extended this structure with attitude in terms of defining two components, media attitude and brand/produce attitude [33]. Accordingly, this theory and its extension provide an overarching theoretical basis to better understand the relationship structure of how advergame influences players' purchase decisions.

However, the hierarchy of effects theory fails to stipulate what a consumer's cognitive beliefs would affect attitude toward brands/products or subsequent purchase intention. Many studies have argued consumer's beliefs as important concerns for using advertising media from an expanded perspective, generally comprising both rational and emotional. The rational belief indicates a design of game for how a brand is placed in the advertising media [37] as the emotional belief shows a psychological state for how a player is motivated with a game context [32].

For advergame design, researchers noted the importance of placing particular forms of brand communication in advertising media for building attitude toward brand, that is, game-brand fit [3]. Furthermore, perceived interactivity is important for the quality of website design as it plays a key role in enabling players to have better communication, self-control of experience, and response to websites [31][34]. Accordingly, we consider game-brand fit and perceived interactivity for this issue. For individual psychological state, prior studies have noted that two psychological states, involvement and escapism, that consumers find, when processing advertising messages, can impact affective state (e.g., brand attitude) [32]. Grounding on the arguments, this study proposes a relationship structure for predicting purchase intention, including four components, belief structure, advergame attitude, brand attitude, and purchase intention.

Moreover, this study considers two control variables, gender and prior experience, to control extraneous sources of variance and to maximize explanatory power. Mehta [26] argued that advertising effectiveness for purchase behavior should allow for gender difference. A person's game experience may influence the link for attitude to purchase intention [11][17]. The following shows the rest of the content. In Section 2, a literature review with hypothesis development for the research model is presented. In Section 3, the related works include scale definition, sampling design, and measurement model. In Section 4: using PLS tool conducts hypothesis testing for estimating path and determination coefficients. In Section 5, the findings are presented with the discussions for the reasons behind them. In Section 6, conclusions are discussed with practical and academic implications.

II. LITERATURE AND HYPOTHESIS

Based on the above discussion, Figure 1 provides a pictorial depiction of this research model. The proposed model is consistent with the basic logic of the hierarchy of effects theory, cognitive, affective, and conative, when often applied in the advertising. Relevant literature and hypotheses are discussed below.

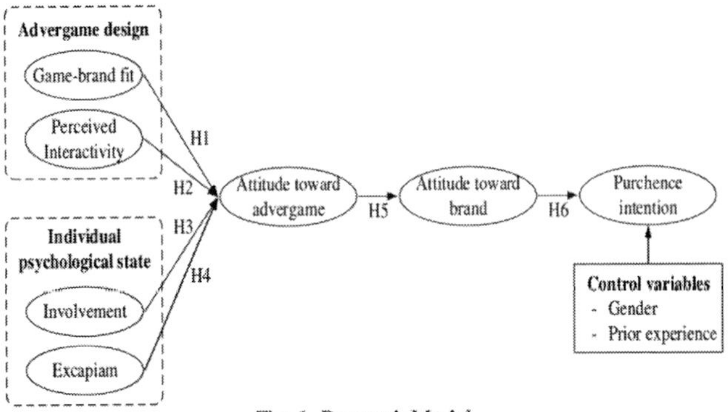

Fig. 1. Research Model

A. *Game-Brand Fit and Advergame Attitude*

Researchers have argued that while playing an online advergame, players would assess whether the nature of the game matches the brand, which tries to relate features of the game to the presence of a brand logo, trademark, product, or spokesperson [27]. A good game-brand fit should generally induce more positive affective responses, such as favorability toward advergame [3]. Gross [11] explored the influence of game-branded product congruity on brand memory and attitude toward branded items placement in games, and reported that this has a significant effect on both brand memory and attitude toward the game. Wise et al. [39] contended that the thematic connection between an advergame and the brand it represents directly affects consumer's information processing, and thus has implications for their attitude toward advergame. Based on this, it is hypothesized:

Hypothesis 1: Game-brand fit positively affects attitude toward advergame.

B. *Perceived Interactivity and Advergame Attitude*

Perceived interactivity is an important determinant for game design as it plays a key role in enabling players to have better communication, self-control, and response with game [35]. Several studies on advertising conducted an experiment by manipulating the design of various interactive levels to customers in order to understand their differences [10]. The results indicated that perceived interactivity is a strong predictor of attitude toward advertising. For website design, researchers attempted to identify the relationship between perceived interactivity and attitude toward the website [20], and they found that perceived interactivity is highly correlated with attitude toward the website. Hence, it is hypothesized:

Hypothesis 2: Interactivity positively affects attitude toward advergame.

C. *Involvement and Advergame Attitude*

Involvement typically amounts to a positive subjective experience/state of an individual toward online activities [32]. When consumer's involvement comes to forming positive attitude toward online advertising, it has a direct impact on attitude toward online advertising [38]. Further studies classified prior empirical studies for finding relationships between involvement and affective responses and user attitude toward IS adoption is affected by the level of involvement [24]. Santosa et al. [32] designed an online information-seeking activity to examine the relationship between personal involvement and user satisfaction and showed that involvement positively affects satisfaction level with this activity. Accordingly, it is hypothesized:

Hypothesis 3: Involvement positively affects attitude toward advergame.

D. *Escapism and Advergame Attitude*

Youn and Lee [40] applied a uses-and-gratifications approach to examine the motivations of playing advergame for college students. Open-ended questions were used to collect a list of motivations for playing advergame online. After completing the frequency analysis, the results showed that escapism is one of the major motivations for playing such games. They also found a strong positive relationship between escapism and attitude toward advergame. Hernandez [12] examined the factors contributing to positive attitude toward advergame among students in Mexico. She also argued that a high level of escapism has a positive relationship with attitude toward advergame, as seen in prior studies. Hence, it is hypothesized:

Hypothesis 4: Escapism positively affects attitude toward advergame.

E. *Advergame Attitude and Brand Attitude*

Researchers on advertising value have demonstrated consistent relationships between attitude toward advertisement and attitude toward brand in both traditional and online advertising [2][8]. Mackenzie et al. [22] also reported that attitude toward advertising is a causal mediating variable in the process through which an advertisement influences attitude toward brand and purchase intention. Wise et al. [39] showed stronger support for the effect of attitude toward advergame on attitude toward brand in a condition of a high thematic connection. Hence, it is hypothesized:

Hypothesis 5: Attitude toward advergame positively affects attitude toward brand.

F. *Brand Attitude and Purchase Intention*

According to the hierarchy of effects theory, cognitive belief precedes affective response (e.g., attitude), which in turn precedes conative behavior (e.g., purchase intention) [36]. Attitude is a predisposition to react favorably or unfavorably toward a particular object [22]. Intention is people's prediction about their own behavior, and is influenced by attitude [28]. Several studies have identified attitude toward brand as an important determinant of purchase intention in the traditional advertising [5][23] and Internet advertising [15][16]. Hence it is hypothesized:

Hypothesis 6: Attitude toward brand positively affects purchase intention.

III. RESEARCH DESIGN

This part is for defining data collection procedure, including scale design, sampling design, and scale validation.

A. *Instrumentation*

The instrument used to gather data was a five-part questionnaire. The first part uses a nominal scale and the others use a 7-point Likert scale.

1) Basic information: We collected basic information about the characteristics of the respondents, including gender, age, education, occupation, types of advergame, number of advergame played before, and frequency of playing advergame.

2) Antecedents of attitude toward advergame: We examined four antecedents of attitude toward advergame. The three items for measuring game-brand fit were adapted from the instruments developed by [13][27]. The three items for measuring perceived interactivity were adapted from the instruments developed by [21][34]. The three items for measuring involvement were adapted from the instruments developed by [32]. The three items for measuring escapism were adapted from those developed by [12][25].

3) Attitude toward advergame: The three items for measuring attitude toward advergame were adapted from the instruments developed by Hernandez [11][12].

4) Attitude toward brand: The three items for measuring attitude toward brand were adapted from the instruments developed by [3].

5) Purchase intention: The three items for measuring purchase intention were adapted from the instruments developed by [18].

B. *Sample Design*

This study primarily explores consumer's purchase intention in terms of the promotional approach of advergame. It would be better to define the qualified respondents in this setting. They were asked to reflect on a recent playing experience of advergame (within the past three months). This manipulation would assure that all respondents are in a consistent manner to recall their experience across all constructs. This manipulation could avoid a bias in the data collection process. Prior research has proposed a similar approach online shopping context [24]. An online survey was placed in online communities to seek players of advergame as potential respondents. Online survey is not also consistent with the context of online game, but also geographically unlimited [7].

A wide variety of data sources was included for survey, allowing the responses to be more representative of the population. Public notices about the survey were posted on a number of bulletin board systems and forums. A reward system was also provided for the respondents. At least 30 participants were drawn from the response sample, with a reward of 5 US dollars being paid to each of them.

C. *Scale Validation*

A pretest was conducted for the scale with practitioners to guarantee acceptable initial reliability and validity. Afterwards, online survey was performed for the above procedure. A total of 453 responses were received with a certain level of experience in playing advergame. After 108 invalid responses with incomplete information were deleted, the final sample size was 345. Common method variance (CMV) was examined with Harman's single factor test [30]. We included all items for analysis to determine whether the majority of variance could be accounted for by one general factor, that is, more than 50% variance accounted for. As no single factor accounted for the bulk of covariance (less than 50% variance accounted for), meaning there was no CMV.

D. *Measurement Model*

This study employed structural equation modeling (SEM) with PLS software to test this proposed model. SEM-PLS allows latent variables to be modeled as either formative or reflective constructs, and places minimal demands on sample size and residual distribution [4]. First, a measurement model was built for scale validation and then a structural model was developed for path analysis.

First, reliability is assessed by the criterion: Cronbach's α larger than 0.7 [4]. Convergent validity is assessed by three criteria: (1) item loading (λ) larger than 0.70, (2) composite construct reliability larger than 0.80, and (3) average variance extracted (AVE) larger than 0.50 [9]. Next, discriminant validity is assessed using the criterion: the square root of AVE for each construct larger than its correlations with all other constructs [9]. All constructs are satisfactory with the criteria of convergent and discriminant validity.

IV. HYPOTHESIS TESTING

Using SEM-PLS tool conducts hypothesis testing. We present the following testing results. First, there is a requirement to estimate path coefficients and statistical significance for this model. Next, determination coefficient (R^2) is computed for endogenous variables to indicate their predictive power. Determination coefficient found in the analysis is similar to the procedure of multiple regression analysis. SEM-PLS does not provide significance tests or interval estimations and a bootstrapping procedure is used with generating 1000 subsamples to estimate path coefficients and their significance. Figure 2 presents the results of the structural model.

For game design aspect, game-brand fit has a positive impact on attitude toward advergame at 0.05 level (β=0.25), supporting Hypothesis 1. Perceived interactivity is a significant predictor of attitude toward advergame at 0.05 level (β=0.23), supporting Hypothesis 2. For psychological state aspect, involvement is important in affecting attitude toward advergame at 0.05 level (β=0.22), while escapism is not significant for its influence (β=-0.05). Hypothesis 3 is supported but Hypothesis 4 is not. Further, the two aspects jointly explain 55% of variance in attitude toward advergame. Next, attitude toward advergame was found to be a notable predictor of attitude toward brand at 0.01 level (β=0.52), supporting Hypothesis 5 with explaining 46% of variance. In turn, attitude toward brand is an important precursor of purchase intention at 0.01 level (β=0.47), supporting Hypothesis 6 with explaining 42% of variance. Finally, regarding the two specified control variables, gender and prior experience are not significantly associated with purchase intention (β=0.01 and 0.01).

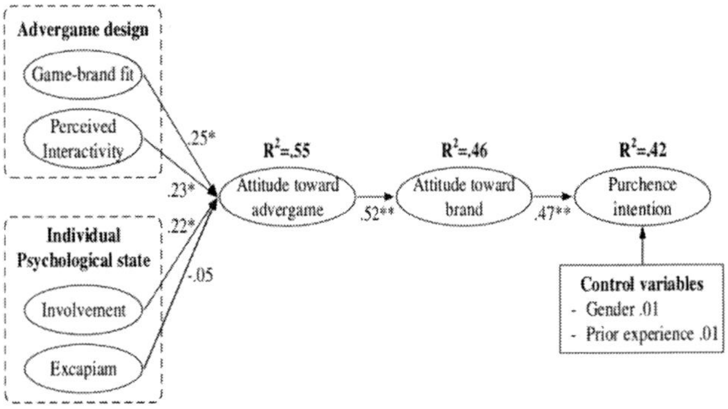

Fig. 2. Structural Model
Value on path: Standardized coefficients,
R^2 : Coefficient of determination, **: p<0.01, *: p<0.05

V. FINDINGS AND DISCUSSIONS

The findings are indicated in Table 1. First, we discuss the possible reasons behind these findings for practice. Next, their importance and contribution could be indicated in literature. In particular, the possible similarities with literature are also discussed.

TABLE I. HYPOTHETICAL TESTING

Hypothetical Links	Results
Game-brand fit->Avergame attitude	Support
Perceived interactivity->Avergame attitude	Support
Involvement->Avergame attitude	Support
Escapism->Avergame attitude	No support
Avergame attitude->Brand attitude	Support
Brand attitude->Purchase intention	Support

A. *Influence of Advergame Design*

Game-brand fit and perceived interactivity indicate the importance in the design of the advergame. Advergame design is rather different from other games in terms of their main purpose and motivation. Advergame is usually initiated by a brand owner, which uses gaming technology to deliver embedded advertising messages to the audience. It attempts to convey brand images and ideas behind a game-playing situation. Thus, when people are playing advergame, they first assess the extent to which brand and game seem to be well matched. If they perceive that brand-game fit is good, they will have positive affective responses [27].

Next, game is in its nature an interactive activity between game and players. Appropriate interactivity as a self-control mechanism can enhance marketing campaigns in an online context [31]. The findings are similar to prior studies in traditional advertising, which indicates that perceived interactivity will influence consumer's attitude toward commercial messages and online advertising [10]. Accordingly, advergame attitude could be positively impacted by game-brand fit and perceived interactivity.

B. *Influence of Psychological State*

When involvement is important for consumer's attitude, this indicates that different people perceive the same advergame differently, and hence respond to it with different affective states. In terms of high- and low-involvement, high-involvement subjects usually believe that the communicated message affects their cognition and attitude, while low-involvement ones do not believe that the communicated message has any important, meaningful, or personally relevant effects. Marketers should place greater emphasis on the issue of involvement to improve player attitudes toward advergame. The role of involvement that we found in this study is consistent with the findings of some prior studies [38].

In contrast, escapism shows an insignificant impact on consumer's attitude and this finding should be quite interesting to marketers. Hernandez [12] concluded that online games allow a player to escape into an alternative reality that may not be stable in all contexts. Advergame neither causes people to put off things they should do, nor makes them get away from their responsibilities. This may be because most people understand that procrastination will only lead to the tasks at hand to be completed later than would be otherwise possible. Playing advergame is thus unlikely to make players feel like that they are in another world, or keep them in a flow state for a long time.

C. *Influence of Advergame and Brand Attitude*

Advergame attitude is useful in evaluating advertising value and is a key precursor of brand attitude when playing an advergame. Advergame attitude is important for advertising value and needs to be carefully nurtured from major concerns of players when playing advergame. For example, when players do not have enough or any affective response to brand, an advergame can be an important medium for them to connect with brand. Marketers must work to make players feel that an advergame is good, likable, and refreshing, and thus lead to them have positive attitude toward brand. The findings are consistent with previous studies, highlighting the fact for an important antecedent for positive brand attitude [22].

Further, brand attitude should be considered as an important affective criterion for consumers when developing an effective advergame mechanism. Branded entertainment mechanisms are highly penetrable with regard to their target audiences, thus enabling them to accept and understand brand advertising, such as advergame. In turn, this would be further reflected in purchase behavior. Marketers should work on attracting players to stay or feel enjoyment with advergame to have their brand seen favorably by the players, and thus increase consumer buying behavior. The findings are consistent with previous studies on traditional advertising [16].

D. *Influence of Control Variables*

Gender and prior experience are not correlated with purchase intention. One ultimate goal of advertising is to persuade the audience to buy branded items. Advertising may be targeted to sex-specific groups, although shoppers are not limited to one gender. Next, consumers have different degrees of playing experience, which may greatly depend on individual traits (e.g., interests, preferences, and needs), even though advergame is becoming more popular and can easily be found online. Accordingly, neither type of purchases is associated with player's prior game-playing experience.

VI. CONCLUSIONS AND SUGGESTIONS

This part shows the research findings for their practical and academic implications as well as the research limitations.

A. *Practical Implications*

In terms of game-brand fit, marketers must better understand brand since each brand has its story, essence, idealized community, and paradox and then achieve effective communication with game developers to ensure that brands and game features have similar images. For perceived interactivity, marketers and game developers should ensure that advergame is enabled by two-way and synchronous communications as this can then increase positive advergame attitude. Regarding psychological state, marketers need to ensure that advergame reflects personal involvement, that is, what consumers perceive to be important and means a lot to them. Moreover, it may be more effective to target an advergame to consumers of a specific background or age, so that it can be made more relevant to them. Such practices will increase the level of personal involvement, and thus improve positive advergame attitude.

B. *Academic Implications*

To the best of our knowledge, few previous studies have investigated consumer's purchase intentions in relation to advergame value. This study has successfully extended the hierarchy of effects theory, as well as provided a better way of exploring advergame value. This particular approach focuses on two aspects, advergame design and psychological state, for driving advergame attitude, brand attitude, and purchase intention.

C. *Research Limitations*

First, there is a wide range of brands, and each brand has a lot of its own advergame. Moreover, in terms of the fields of applications, advergame can be used by marketers with various branded products and services. However, this study did not focus on any specific brand (e.g., McDonald, Nissan), any specific field (e.g., food, car), and any specific appeals (e.g., humor vs. fear). Hence, these concerns may have affected the generalization of the results and have made it hard to identify different routes to attitudinal change. Next, according to basic demographic information, the majority of respondents are students. However, research has shown that students are the chief players of advergame and are mostly trained to value rationalism, and as a result, may reject an emotional appeal more than other populations. Thus, this group is highly representative of the target population.

D. *Future research*

Further research can be extended to include content design for an entertainment purpose to increase flow state of consumers. Next, a concern of cross-cultural differences may be included to understand consumer behavior. Further, personal IT skills is an important determinant for consumers to be willing to involve in this game playing since it is created by an IT-based form.

REFERENCES

[1] T. E. Barry, "In defense of the hierarchy of effects: A rejoinder to Weilbacher," *Journal of Advertising Research*, vol. 42, no. 3, pp. 44-47, 2002.

[2] I. Buil, L. de Chernatony, and E. Martínez, "Examining the role of advertising and sales promotions in brand equity creation," *Journal of Business Research*, vol. 66, no. 1, pp. 115-122, 2013.

[3] V. Cauberghe and P. D. Pelsmacker, "Advergames: The impact of brand prominence and game repetition on brand responses," *Journal of Advertising*, vol. 39, no. 1, pp. 5-18, 2010.

[4] W. W. Chin, B. L. Marcolin, and P. R. Newstead, "A partial least squares latent variable modeling approach for measuring interaction effects: Results from a Monte Carlo simulation study and an electronic-mail emotion/adoption study," *Information Systems Research*, vol. 14, no. 2, pp. 189-217, 2003.

[5] S. Chow, R. L. Rose, and D. G. Clarke, "Sequence: Structural equations estimation of new copy effectiveness," *Journal of Advertising Research*, vol. 32, no. 4, pp. 60-72, 1992.

[6] J. Culp, R. A. Bell, and D. Cassady, "Characteristics of food industry web sites and advergames targeting children," *Journal of Nutrition Education and Behavior*, vol. 42, no. 3, pp.197-201, 2010.

[7] M. Denscombe, "Web-based questionnaires and the mode effect: An evaluation based on completion rates and data contents of near-identical questionnaires delivered in different modes," *Social Science Computer Review*, vol. 24, no. 2, pp. 246-254, 2006.

[8] C. M. Derbaix, "The impact of affective reactions on attitudes toward the advertisement and the brand: A step toward ecological validity," *Journal of Marketing Research*, vol. 32, no. 4, pp. 470-479, 1995.

[9] C. Fornell, and D. F. Larcker, "Structural equation models with unobservable variables and measurement error: Algebra and statistics," *Journal of Marketing Research*, vol. 18, no. 3, pp. 382-388, 1981.

[10] Q. Gao, P. L. Rau, and G. Salvendy, "Perception of interactivity: Affects of four key variables in mobile advertising," *International Journal of Human-Computer Interaction*, vol. 25, no. 6, pp. 479-505, 2009.

[11] M. L. Gross, "Advergames and the effects of game-product congruity," *Computers in Human Behavior*, vol. 26, no. 6, pp. 1259-1265, 2010.

[12] M. D. Hernandez, "Determinants of children's attitudes towards advergames: The case of Mexico," *Young Consumers*, vol. 9, no. 2, pp. 112-120, 2008.

[13] M. D. Hernandez, and M. S. Minor, "Investigating the effect of arousal on brand memory in advergames: Comparing qualitative and quantitative approaches," *Qualitative Market Research: An International Journal*, vol. 14, no. 2, pp. 207-217, 2011.

[14] Á. Hofmeister-Tóth, and P. Nagy, "The content analysis of advergames in Hungary," *Qualitative Market Research: An International Journal*, vol. 14, no. 3, pp. 289-303, 2011.

[15] C. D. Hopkins, M. A. Raymond, and A. Mitra, "Consumer responses to perceived telepresence in the online advertising environment: The moderating role of involvement," *Marketing Theory*, vol. 4, no. 1-2, pp. 137-162, 2004.

[16] J. Hwang, Y. S. Yoon, and N. H. Park, "Structural effects of cognitive and affective responses to web advertisements, website and brand attitudes, and purchase intentions: The case of casual-dining restaurants," *International Journal of Hospitality Management*, vol. 30, no. 4, pp. 897-907, 2011.

[17] P. Ing, "The impact of advertising position and games experience on purchase intention in advergaming," *Interdisciplinary Journal of Contemporary Research in Business*, vol. 1, no. 4, pp. 40-53, 2009.

[18] M. Lai, and Y. S. Huang, "Can learning theoretical approaches illuminate the ways in which advertising games effect attitude, recall, and purchase intention," *International Journal of Electronic Business Management*, vol. 9, no. 4, pp. 368-380, 2011.

[19] R. J. Lavidge, and G. A. Steiner, "A model for predictive measurements of advertising effectiveness," *Journal of Marketing*, vol. 25, no. 6, pp. 59-62, 1961.

[20] J. Lee, and W. N. Lee, "Antecedents and consequences of perceived interactivity: An exploratory study," *Journal of Interactive Advertising*, vol. 3, no. 1, pp. 34-45, 2002.

[21] Y. Liu, and L. J. Shrum, "What is interactivity and is it always such a good thing? implications of definition, person, and situation for the influence of interactivity on advertising effectiveness," *Journal of Advertising*, vol. 31, no. 4, pp. 53-64, 2002.

[22] S. B. Mackenzie, R. L. Lutz , and G. E. Belch, "The role of attitude toward the ad as a mediator of advertising effectiveness: A test of competing explanations," *Journal of Marketing Research*, vol. 23, no. 2, pp. 130-143, 1986.

[23] S. B. Mackenzie, and R. A. Spreng, "How does motivation moderate the impact of central and peripheral processing on brand attitudes and intentions?," *Journal of Consumer Research*, vol. 18, no. 4, pp. 519-529, 1992.

[24] M. A. Mahmood, J. M. Burn, L. A. Gemoets, and C. Jacqqez, "Variables affecting information technology end-user satisfaction: A meta-analysis of the empirical literature," *International Journal of Human-Computer Studies*, vol. 52, no. 4, pp. 751-771, 2000.

[25] C. Mathwick, "Understanding the online consumer: A typology of online relational norms and behavior," *Journal of Interactive Marketing*, vol. 16, no. 1, pp. 40-55, 2002.

[26] A. Mehta, "Advertising attitudes and advertising effectiveness," *Journal of Advertising Research*, vol. 40, no. 3, pp. 67-72, 2000.

[27] S. Okazaki, and M. J. Yagüe, "Responses to an advergaming campaign on a mobile social networking site: An initial research report," *Computers in Human Behavior*, vol. 28, no. 1, pp. 78-86, 2012.

[28] R. E. Petty, R. H. Unnava, and A. J. Strrathman, "Theories of attitude change," *Handbook of Consumer Behavior*, pp. 241-279, 1991.

[29] C. Pizzutti & D. Fernandes, "Effect of recovery efforts on

consumer trust and loyalty in e-tail: A contingency model," *International Journal of Electronic Commerce,* vol. 14, no. 4, pp. 127-160, 2010.

[30] P. M. Podsakoff, S. B. Mackenzie, J. Y. Lee, and N. P. Podsakoff, "Common method biases in behavioral research: A critical review of the literature and recommended remedies," *Journal of Applied Psychology*, vol. 88, no. 5, pp. 879-903, 2003.

[31] G. Rosenkrans, "Maximizing user interactivity through banner Ad design," *Journal of Promotion Management*, vol. 16, no. 3, pp. 265-287, 2010.

[32] P. I. Santosa, K. K. Wei, and H. C. Chan, "User involvement and user satisfaction with information-seeking activity," *European Journal of Information Systems*, vol. 14, no. 4, pp. 361-370, 2005.

[33] B. M. Sukoco, and W. Y. Wu, "The effects of advergames on consumer telepresence and attitudes: A comparison of products with search and experience attributes." *Expert Systems with Applications*, vol. 38, no. 6, pp. 7396-7406, 2011.

[34] S. S. Sundar, and J. Kim, "Interactivity and persuasion: Influencing attitudes with information and involvement," *Journal of interactive advertising*, vol. 5, no. 2, pp. 5-18, 2005.

[35] C. Tangmanee, and J. Rustanavibul, "Product placement in online games: An exploration into Thai gamers' attitude," *The Journal of American Academy of Business, Cambridge*, vol. 17, no. 2, pp. 92-98, 2012.

[36] D. Vakratsas, and T. Ambler, "How advertising works: What do we really know?." *Journal of Marketing*, vol. 63, no. 1, pp. 26-43, 1999.

[37] E. A. van Reijmersdal, J. Jansz, O. Peters, and G. Van Noort, "The effects of interactive brand placements in online games on children's cognitive, affective, and conative brand responses," *Computers in Human Behavior*, vol. 26, no. 6, pp. 1787-1794, 2010.

[38] E. Wang, "Forming positive advertising and product attitude: The role of product involvement," *International Journal of Technology Marketing*, vol. 6, no. 3, pp. 259-271, 2011.

[39] K. Wise, P. Bolls, H. Kim, "Venkataraman, A., and Meyer, R. Enjoyment of advergames and brand attitudes: The impact of thematic relevance," *Journal of Interactive Advertising*, vol. 9, no. 1, pp. 27-36, 2008.

[40] S. Youn, and M. Lee, "Advergame playing motivations and effectiveness: A uses and gratifications perspective," *In: Advertising, Promotion, and New Media*, M. R. Stafford and R. J. Faber (eds.), New York: Sharpe, pp. 320-347, 2005.

Mobile Enterprise Application Development in Practice: an Analysis of Real-World User Stories

Matthias Jurisch, Stephan Böhm, Toby James-Schulz
Faculty of Design – Computer Science – Media
RheinMain University of Applied Sciences
Wiesbaden, Germany
Email: {matthias.jurisch, stephan.boehm}@hs-rm.de,
t.jamesschulz@outlook.com

***Abstract*—In development of mobile enterprise applications, saving time is an important factor. Short time to market and always changing technical environments require the ability to adapt to changes. However, these demands are in conflict with writing modular, well-thought-out software that allows easy reuse from results of past projects where requirements are often documented in user stories. These short descriptions of ways for a user to interact with the system are often the center of application development. In this paper, we present an analysis of user stories in practice and a categorization of terms used in these stories. We also evaluate how these categorizations relate to algorithmic similarities from information retrieval.**

Keywords–User Story; Mobile Enterprise Applications; Clustering

I. Introduction

In recent years, user centered design approaches have proven to be beneficial for software and mobile application(app) development. User centered design consists of a variety of tools to assess and evaluate user requirements during the development process, with the goal of delivering software up to par with the users usually high usability requirements. The assessment of user requirements at the beginning of the development process saves time and resources, as focusing on implementing features the user actually needs, leads to less changes having to be made in later stages of development. A way of gathering information about the requirements is through user stories [1], in which the requirements are described from a user's point of view. These stories then serve as a foundation for the further development of source code, screen designs and other software artifacts. Involving the user in the development process is not the only way enterprises can save resources during development, especially larger enterprises can save a lot of time through properly reusing existing software artifacts. Alas there should be a focus on standardized documentation to support the reuse of these software artifacts. This is a challenge large enterprises face, especially with multiple development projects running at the same time. Due to the fast pace of development, the developers often do not have time left for meaningful post-processing of software artifacts, leading to a lack of documentation and documentation standards.

Setting standards and building a library of well documented software artifacts in hindsight, forces a company to allocate resources for theoretically already finished development projects. This would lead to manpower being occupied with working off backlogs, instead of working on the latest projects. Saving time and resources through reuse is a desirable goal for organizations in the Mobile Enterprise Application (MEA) [2] market, but the quick time to market and the fast paced nature of the market environment leave no spare time to deal with backlogs, as new issues and tasks arise on a daily basis. This calls for automated methods to relate stories to each other, which can be built on well-known information retrieval methods. Automatically connecting user stories and recommending them to a developer working on a story would have several benefits: developers might not be aware that similar stories might exist and therefore will not search for similar stories on their own. Also, developers do not need to come up with search terms if they want to find similar stories. When looking at similar stories, developers can reuse existing source code or other artifacts for implementing the story they are currently working on. Computing the similarity of stories would also allow directly recommending other artifacts such as source code or screen designs. However, existing methods for relating user stories to each other often fall short on identifying synonyms and accurately representing domain-specific vocabulary.

To overcome this issue, in this work, we manually analyze a set of real-world user stories regarding the vocabulary used in the most relevant parts of the user story. The results of this analysis are then used to automatically categorize a larger set of user stories from real-world mobile enterprise application development projects. We also compare this automatic categorization to a standard similarity measure from the area of information retrieval.

The remainder of this work is structured as follows: Section II introduces the overall setting of mobile enterprise application development and gives some background on user stories. Section III identifies related work and a research gap that we address in this paper. Our approach is presented in Section IV. In Section V, we introduce our dataset. Term clustering results are shown in Section VI. Results of using these term clusters to categorize stories are discussed in Section VII. The relationship between these clusters and similarity measures from information retrieval is analyzed in Section VIII. Section IX discusses practical implications of this work. A conclusion is given in Section X.

II. Background

Mobile Enterprise Applications (MEA) is not a term with an exact definition [2]. In this work, we use this term to

```
As a <user>
I want <feature>
So that <reason>
```

Figure 1. User Story Template

describe applications that are created or used in the context of the daily work of enterprises. These mobile enterprise applications are, just like regular mobile applications, often developed based on user stories. In modern software development, user stories are a common tool to manage and document user requirements. A user story should describe, what kind of interactions a user wants a software system to support and how this is beneficial for the user. The most common template for this is the role-feature-reason or Connextra-format [1]. This template is shown in Figure 1.

The *user* aspect of this template can relate to several aspects. Organizational roles as well as platforms (e.g., "Tablet User") can be used. The *feature* aspect represents the kind of interaction the user wishes a system to support and the *reason* represents the reason, why a user needs this kind of interaction. A typical example for a user story is: *As a user I want to mark and select favorites in order to receive information about my daily bus and train connections as fast as possible.* User stories are often accompanied by *acceptance criteria*, that define required properties of the implementation of a user story. Furthermore, there are several guidelines for creating user stories, one of these are the INVEST criteria [3]. These criteria state, that a user story should be *independent* from other user stories, *negotiable*, *valuable* with a benefit for the user that is clearly identifiable, *estimable* regarding its cost, *small*, and *testable* or verifiable.

In practice, the quality and granularity of user stories may vary. In our experience, the *reason* aspect of user stories is often left out. Besides, user stories are not always formulated in a way that they are easily understandable for an outsider. When trying to use user stories to improve software reuse, this is an important challenge.

III. Related Work

Using short descriptions, such as user stories to support software reuse is not an entirely new idea. Earlier works in this area are based on bug reports. Hipikat [4] proposes using bug reports, which are also short textual descriptions and contain some amount of information about a requirement or software change request, to build a system for recommending software artifacts that can be reused. This system is based on textual similarity of issue descriptions using information retrieval techniques. A related area is the area of issue triage, where software systems recommend developers for a given issue based on the history of a project. In this area, many approaches use methods from the area of information retrieval [5][6]. More recent approaches in this area use deep learning methods [7].

While these approaches have been applied to bug reports, only few approaches have applied these ideas to user stories: [8] proposes a recommendation system based on user stories and evaluates this system on a project history of a single project. Hence, no inter-project knowledge exchange is examined. In our previous work [9], we evaluated how well information-retrieval-based approaches can distinguish between two types of user stories and which aspects of the user story are important to it and collected first evidence on how these approaches perform on real-world data [10]. However, some important properties of user stories in the context of mobile development are unclear: First of all, quality and adherence to the structure of user stories used in practice in this area have not been assessed in the literature. Hence, it is not clear to what degree these user stories can be used for building a recommendation systems that suits practical needs and what kind of semantic similarities can be discovered in this kind of data. Also, using information-retrieval-based similarity measures often have the drawback that synonyms or semantically similar terms can not be identified. While the impacts of this could be offset by using term representations that encode semantic of terms like word embeddings, it is very common in this context to use proprietary or enterprise-specific terms that are not easily represented using these methods.

In this work, we tackle these issues by analysing a set of user stories from real-world mobile enterprise application development projects. Results of these analysis are then used as an input for clustering stories into different categories based on the terms used in specific parts of stories.

IV. Approach

In this work, we want to answer the following research questions:

1) How are user stories used in the mobile enterprise application development context? What parts of user stories are more/less important to developers?
2) Can terms in specific parts of the user story (e.g., in the user or feature part) be separated into clusters based on their semantics? If this is the case, what are the properties of these clusters?
3) How are these clusters related to similarities based on information retrieval techniques?

To address these questions, we first analyze a set of real-world user stories by hand. This manual annotation process is depicted in Figure 2. First, a data cleaning step is required (1). In this step we only select issues that contain text that can be identified as a user story. In the next step (2) we split these issues into acceptance criteria, story title and actual content. In the actual story content is split up into the user, feature and rationale content (3). The users are then clustered into different user types (4).

For the feature part, some more steps for a meaningful clustering are required. When looking at our data we found that many feature descriptions can be summarized by a verb and an object (e.g., "see search results", "enter address" or "edit favorites"). Hence, in one step we select these key verbs and objects (5). Both verbs (6) and objects(7) are then clustered in the last step.

These clusters of elements of user stories can then be used to separate the user stories themselves into three types of clusters (one for each user, one for each key feature verb and one for each key feature object). With the set of terms for each cluster, it is possible to search for these terms in the user stories and assign them to the respective cluster if they contain a term from it.

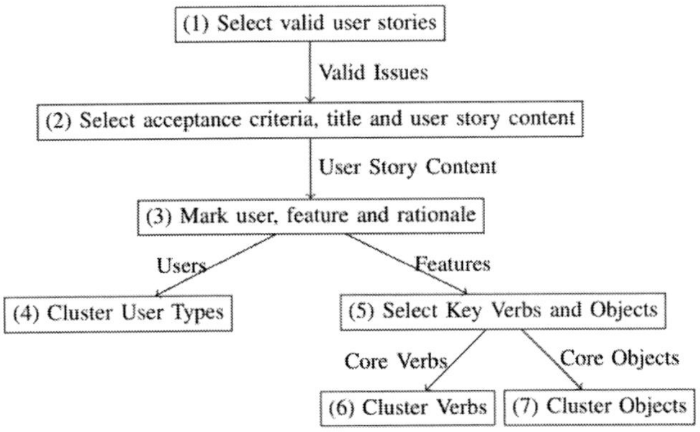

Figure 2. Manual clustering process

These clusters can then be examined regarding their distribution – if all user stories are basically in the same cluster, the clustering method is not very meaningful, while if the distribution is more even among clusters, these clusters might contain some more information. This evaluation is supposed to provide some insights on the second research question.

To answer our third research question, we will compare the similarities of stories inside a cluster with the similarities between stories of the overall dataset. We will use the similarity measures used in [10] to compute the similarities. Namely, we will use TF-IDF-based similarities. TF-IDF is a method for document representation based on term occurrence in documents that is very common in information retrieval [11]. Each document d (i.e., a user story) is represented by a vector $\boldsymbol{W}_d$, that contains an entry for each term used in the dataset. Each vector component $\boldsymbol{W}_{d,t}$ represents the importance of a term t for the document d.

This importance is computed by multiplying $tf_{d,t}$,the frequency of term t in document d, by a representation of how common the term t is in all documents. For measuring the commonality of the t ,the inverse document frequency $\log \frac{N}{df_t}$ is used. N represents the number of all documents and df_t is the number of occurrences for t term in all documents. This yields the following formula for a document's vector representation:

$$\boldsymbol{W}_{d,t} = tf_{d,t} * \log \frac{N}{df_t}$$

To compute the similarity of two documents d_1 and d_2, the cosine of the angle between their vector representation is used.

V. Dataset

The dataset of our evaluations consists of 1408 issues from a Jira system that are not necessarily labeled as *User Stories*. The Jira system is used for organizing app development in a department that mainly focuses on the development of mobile enterprise applications. The Jira System is used by around 100 Users. The user stories stem from more than 20 development projects, where a project is usually implemented by a small team using an agile development approach. An average project has around 50-100 user stories associated to it. While all projects stem from the same company, projects vary in size and type. Also, several different teams have worked on the projects examined in this paper.

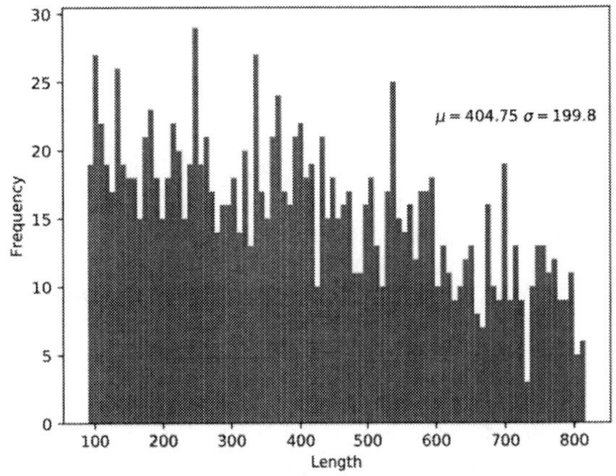

Figure 3. Histogram of User Story length

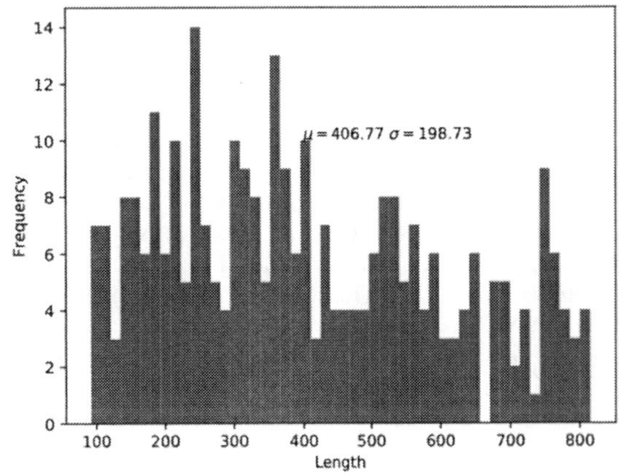

Figure 4. Histogram of User Story length of subset

As in [10], we only considered user stories that contain at least 80 characters of text, since the template alone already contains around 40 characters. The average length of these stories is 404.75 characters with a standard deviation of 199.8. A histogram of story length is displayed in Figure 3. This highlights an important aspect of user stories in practice: there is a large variety in the content and structure of user stories.

Since manually labeling around 1500 stories would be very time consuming, we created a sub-sample of 300 randomly selected stories for the labeling process. Regarding length statistics, this sample is very similar to the full dataset. Mean length is 406.77 characters and standard deviation 198.7. A histogram of these lengths is shown in Figure 4. The average length of text that actually describes the user story (without acceptance criteria, title, etc.) is 99.30 characters. The mean length of the user is 12.81 characters, the mean feature Length is 60.73 characters and the mean rationale length is 47.47

characters. The overall average text length of issues that contain stories is 458.612 characters. All stories consist of a user and a feature, whereas only 58.82% contain a rationale.

VI. Term Clustering

For finding clusters in this dataset, we followed the process described in Section IV. After the first data cleaning step, where we only selected issues that actually contain a user story, 132 issues remained. In these stories, we searched for three types of clusters: (1) Clusters of *User Types*, (2) clusters of *core feature verbs* and (3) clusters of *core feature objects*.

For *users*, we found six clusters. The first of these group consists of several synonyms and translations for the word "user". Another group of users are user descriptions with a platform specification (e.g., "Tablet User"). A third and largest group contains user descriptions that are in some way related to a role in an organization (e.g., "developer"). Another group also relates to a role, while these users are associated with some kind of privileged roles (e.g., "admin, supervisor"). The fifth group is made up of external users such as customers. The sixth group includes users where a broader term, such as the whole department was used, e.g., quality control.

For *core feature verbs*, we found eight clusters. The first three of these clusters are comprised of terms for creating, updating and deleting data. The fourth cluster contains terms related to viewing and working with collections of data, such as sorting or filtering. The fifth cluster of terms is comprised of data management features related to exporting, sharing and importing data. The sixth cluster is related to system management features such as user management and notifications. Another cluster is dedicated specifically to search functionalities. The eighth group of terms is comprised of vocabulary for interacting with conversational interfaces, such as greeting and talking.

For *core feature objects*, we found seven clusters. The first cluster contains widget names. The second cluster groups terms for several types of data such as records or documents. Another cluster contains technical terms such as backend or platform. The fourth cluster represents feedback options for users of the applications. A fifth cluster is related to error handling. The sixth cluster contains terms for an app overview. The seventh cluster groups terms for notifications.

VII. Story Clustering

As described in Section IV, we used the clustered terms for sorting user stories into clusters, based on the existence of the clustered terms in the stories. For *user types*, this lead to cluster sizes as depicted in Table I. Clusters are listed in the same order as they are introduced in Section VI. Note that while the clusters for terms can not overlap, the clusters of user stories based on these terms can, since stories may contain terms from several clusters. The first conclusion from these clusters is, that the term cluster for *Department* seems to not be very important for most user stories, since we can find only 4 stories that contain terms from this cluster. As expected, the cluster containing synonyms for a *Generic User* leads to many user stories. More than a third of stories are part of this cluster.

Feature clusters based on verbs are shown in Table II. Clusters based on core feature verbs seem to be more balanced than clusters based on user terms. Cluster sizes range from 75-313 stories. Hence, these clusters are more likely able to divide the dataset into more meaningfully separated groups than clusters based on user terms.

TABLE I. CLUSTERS BASED ON USER TERM CLUSTERS

Cluster	*Number of Stories*
Generic User	570
Platform	128
Organizational Role	159
Supervisor Role	130
External Users	146
Department	4

TABLE II. CLUSTERS BASED ON CORE FEATURE VERBS

Cluster	*Number of Stories*
Create	202
Edit	160
Delete	75
Collections of Data	313
Importing/Exporting	250
System management	233
Searching	142
Conversational	296

TABLE III. CLUSTERS BASED ON CORE FEATURE OBJECTS

Cluster	*Number of Stories*
Widgets	363
Documents and Reporting	643
Technical Terms	679
Feedback	163
Error Handling	16
Views and Presentation	87
Notifications	66

Story clusters based on core feature objects are shown in Table III. There are two large clusters, namely *Record Types* and *Technical Terms* that contain many user stories and are hence probably to general to be used for separating the data. The cluster for *Error Handling* seems very small, while other clusters are in a similar range to clusters based on feature verbs.

VIII. In-Cluster Similarities

To evaluate how these manually determined clusters relate to methods from information retrieval, we computed the similarity values of the 5 most similar stories that are in the same cluster and the similarities of all stories to the top 5 most similar stories regardless of cluster. We chose the top five most similar stories, since these are the stories that would be on top of a recommendation list. A histogram of similarities is shown in Figure 5. The mean for in-cluster similarities is 0.2859 and the standard deviation is 0.2056. The mean for similarities of all stories is 0.3032 and the standard deviation is 0.2008.

Contrary to our intuition, the distributions of similarities are fairly similar, with the mean for all similarities marginally higher than for similarities in the same cluster. This means that the most similar stories are very often, but not always in the same cluster. Hence, these clusters carry similar information to what is encoded with information retrieval techniques.

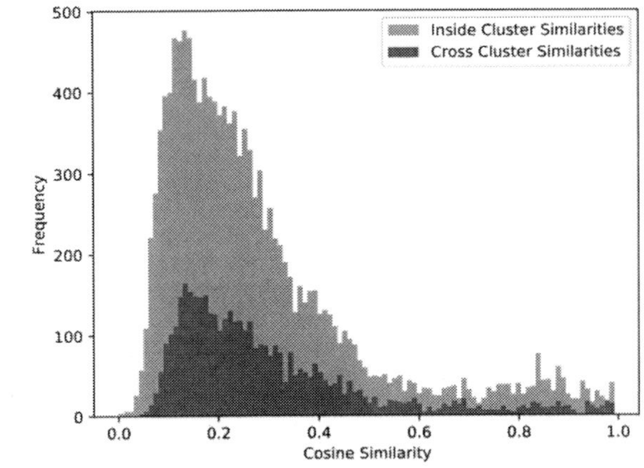

Figure 5. Histogram of top 5 User Story Similarity Inside Clusters compared to Story Similarity of all Stories

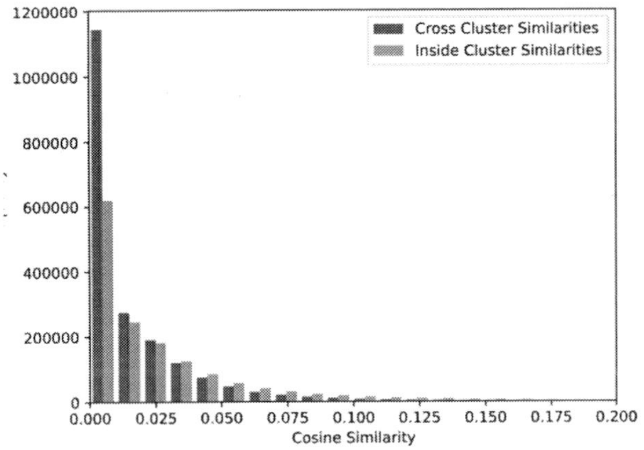

Figure 6. Histogram of User Story Similarity Inside Clusters compared to Story Similarity of all Stories

To further investigate this issue, we also computed the similarities of all stories in the same cluster and the similarities of all issues regardless of cluster. A histogram for both scores is given in Figure 6. The histogram shows only similarities between 0 and 0.2, since higher scores are relatively rare in comparison to this interval. Mean similarities inside clusters is 0.0302 with std. deviation of 0.0523. Mean similarity of all stories is 0.0172 wit std. deviation of 0.0342. This shows that clusters especially filter stories that are dissimilar according to TF-IDF.

IX. Practical Implications and Discussion

Regarding our research questions formulated in Section IV, we can draw the following implications. User Stories are an important part of daily work in mobile enterprise application development - we found that a significant amount of stories exist in a real world dataset. We also found that the rationale part of stories is frequently left out, which is not the case for other parts of the template. The rationale for leaving out the *reason* can be based on two factors: either the *reason* is woven into the *feature* part of the user story, or the author of the story sees the *reason* based in common sense. An example for this might be "as a user I want a cancel button" or "as a user I want to receive error notifications". This leads to the longest and most detailed part of a story being the feature part. Another finding is that a generic user description (e.g., "User of the application") is the most common form of user type. User stories have the benefit of delivering a lot of information with little text. A practical implication arising from this is that further education on proper use of user stories is necessary to optimize information retrievability not only for humans but also for algorithms. The same result could probably be achieved by streamlining the templates of the Jira System, to make the resulting user stories more consistent in their form. Another implication is that it could be beneficial to create a repository with user stories describing basic features that every project needs. The repository should give the developer an easy overview of which features are missing e.g., can the user cancel input without losing data?, does the user receive easy to understand error messages?, etc. A repository like this could even be integrated into a templating system to automatically generate stories when a project is created.

Clustering terms into different types of categories for users, feature verbs and feature objects also led to some insights. The clusters for users show, that a large portion of user roles in Mobile Enterprise Application Development are related to organizational aspects. Another aspect are platforms, on which the application is run, but the most common roles are organizational or generic. From building clusters for feature terms, we found that classical paradigms such as the CRUD-pattern [12] can be used to categorize terms used in user stories. Also, many stories consider some type of handling lists or collections of data and system management as well as transferring data. A different area is related to conversational interfaces. Feature objects come from several different categories: They can be related to display items such as widgets or can relate to technical terms for application components.

When it comes to improving recommendations through these clusters, we found that clusters based on our manual categorization and similarities computed using information retrieval methods overlap in many cases – TF-IDF-based similarities in the same cluster are nearly identical to similarities of stories in general when only considering the top 5 most similar stories. This indicates that using these clusters is not likely to be a successful way of improving similarity measures.

X. Conclusion and Outlook

In this paper, we presented an analysis of user stories used in a real-word mobile enterprise application development context. Our main finding is, that user stories in this domain can be categorized based on specific parts of the common template. These categorizations contain similar information to what can be achieved through automatic information retrieval methods. We also found, that the rationale part of user stories is frequently (in roughly 40% of stories in our dataset) left out while other aspects like the feature description seem to carry more meaning for story authors.

As future work several directions are possible. A comparative evaluation of stories from consumer application development and mobile enterprise application development could help to highlight differences and challenges that are important in both sectors. Another possible aspect of future work might be creating a set of synonyms or even an ontology of enterprise-specific terms to improve similarity measures.

ACKNOWLEDGMENT

This work was funded by the German Federal Ministry of Education and Research, grant no. 03FH032PX5; the PROFRAME project at RheinMain University of Applied Sciences. All responsibility for the content of this paper lies with the authors.

REFERENCES

[1] M. Cohn, User Stories Applied: For Agile Software Development. Redwood City, CA, USA: Addison Wesley Longman Publishing Co., Inc., 2004.

[2] A. Giessmann, K. Stanoevska-Slabeva, and B. de Visser, "Mobile enterprise applications–current state and future directions," in System Science (HICSS), 2012 45th Hawaii International Conference on, Jan 2012, pp. 1363–1372.

[3] B. Wake, "INVEST in good stories, and SMART tasks," blog post, [retrieved: 2019.09.10], 2003. [Online]. Available: http://xp123.com/articles/invest-in-good-stories-and-smart-tasks

[4] D. Cubranic, G. C. Murphy, J. Singer, and K. S. Booth, "Hipikat: A project memory for software development," IEEE Trans. Softw. Eng., vol. 31, no. 6, Jun. 2005, pp. 446–465.

[5] J. Anvik and G. C. Murphy, "Reducing the Effort of Bug Report Triage: Recommenders for Development-Oriented Decisions," ACM Transactions on Software Engineering and Methodology (TOSEM), vol. 20, no. 3, 2011, pp. 10:1–10:35.

[6] P. Runeson, M. Alexandersson, and O. Nyholm, "Detection of duplicate defect reports using natural language processing," Proceedings - International Conference on Software Engineering, 2007, pp. 499–508.

[7] S. Mani, A. Sankaran, and R. Aralikatte, "Deeptriage: Exploring the effectiveness of deep learning for bug triaging," arXiv preprint arXiv:1801.01275, 2018.

[8] H. Pirzadeh, A. D. S. Oliveira, and S. Shanian, "ReUse : A Recommendation System for Implementing User Stories," in International Conference on Software Engineering Advances, 2016, pp. 149–153.

[9] M. Jurisch, M. Lusky, B. Igler, and S. Böhm, "Evaluating a recommendation system for user stories in mobile enterprise application development," International Journal On Advances in Intelligent Systems, vol. 10, no. 1 and 2, 2017, pp. 40–47.

[10] M. Lusky, M. Jurisch, S. Böhm, and K. Kahlcke, "Evaluating a User Story Based Recommendation System for Supporting Development Processes in Large Enterprises," in CENTRIC 2018, The Eleventh International Conference on Advances in Human-oriented and Personalized Mechanisms, Technologies, and Services, 2018, pp. 14–18.

[11] D. Jurafsky and J. H. Martin, Speech and Language Processing (2Nd Edition). Upper Saddle River, NJ, USA: Prentice-Hall, Inc., 2009.

[12] J. Martin, Managing the Data Base Environment, 1st ed. Upper Saddle River, NJ, USA: Prentice Hall PTR, 1983.

Motivational Features in an Application for Presenting Dysfunctional Movement Patterns and for Providing Support in Conducting Exercises

Marie Sjölinder, Anneli Avatare Nöu
RISE, Research institutes of Sweden
Stockholm, Sweden
e-mail: marie.sjolinder@ri.se, anneli.nou@ri.se

Vasiliki Mylonopoulou, Olli Korhonen
University of Oulu
Oulu, Finland
e-mail: Vasiliki.Mylo@oulu.fi, Olli.Korhonen@oulu.fi

Abstract— **This paper describes motivational features in a mobile application for physiotherapy related exercises. The features support goal setting, possibilities to follow progress, personalization and possibilities to compare own progress or performance with other users. During the iterative development of the application, an explorative study was conducted where the participants were interviewed about the motivational aspects described above. The respondents emphasized the importance of goal setting together with the physiotherapist and being able to follow progress. With respect to being able to compare performance or progress with other users, the outcome of our work is in line with previous research where comparisons have been rejected. However, people may be more positive if the comparison is disguised as a gamification element, for example, as a part of a competition or in terms of contributing to a group.**

Keywords-movement related disorders; mobile application for conduction exercises; motivational theories; social comparison; personalization.

I. INTRODUCTION

Movement related disorders is a common occupational disease in the European Union and workers in all sectors and occupations are affected [1]. This is an increasing problem and one of the most important causes of long-term sickness absences. Early detection and early intervention could reduce the number of serious movement related problems. By gathering and analyzing movement data from large groups of people over a long period of time, categorization of different movement related patterns can be made. Based on this categorization, one person's movement pattern can be placed into one cluster and early signs of problems and movement related disorders can be detected before it has started to cause problems or pain. Depending on this knowledge, relevant and individualized support and exercises can be suggested using smartphone applications. However, the challenge is to motivate the users to conduct the suggested exercises based on individual recommendations from the physiotherapist, and to comply with training programs aimed at solving possible future problems.

In this work motivational features were applied in a mobile application for physiotherapy related exercises. The features were related to goal setting, providing support in follow progress, personalization and possibilities to compare own performance with others. Interviews and gaining feed-back from users were conducted as a part of a larger work where the application was developed in an iterative way with different user groups. The aim with interviews was to gain a deeper understanding of how to apply motivational features and personalization when developing applications based on large amounts of aggregated movement related data. The work does not claim to investigate different motivational models in a systematic way. Instead, it was an explorative study highlighting the use of different social motivational aspects in developing an application providing support in conducting exercises. In the following text, section 2 describes the project and the concept that the developed application was a part of. Section 3 gives a short overview of motivational theories. The categories of motivational theories and the concepts that are described are central and discussed in terms of possibilities to be applied in the context of the developed application. Section 4 presents the outcome of the explorative study that was conducted as a part of the iterative development. Based on the study, Section 5 suggests design implications and also gives examples of how these features were implemented within the developed application. Finally, section 6 discusses the work conducted and suggests possible future work.

II. AN APPLICATION FOR SUPPORTING PHYSIOTHERAPY RELATED EXERCISES

The application, developed as a part of this work, was based on the company Qinematic's software service that record and analyze body movements using 3D digital video. The users are standing in front of a Kinect sensor and follow instructions about movements to conduct. Based on these sessions, 3D-data is gathered and stored. As an extension to this service a research project was conducted that had two aims. The first aim was to develop machine learning algorithms to analyze gathered movement data, and the second aim was to develop user applications to provide information about dysfunctional movement patterns, facilitate contact with healthcare providers, make it possible for physiotherapists to suggest exercises and for the users to set goals and follow their progress (Figure 1). Via the application, the healthcare provider also had the possibility to gather further information by asking health related questions to the users. This was done to offer a better and more personalized care. As mentioned, the entire concept/system consisted of several parts, including machine learning and categorization of dysfunctional movement patterns. The work presented here focuses on the development of motivational features in the application targeted towards users with possible dysfunctional movement patterns. However, the larger concept around the application placed other demands related to motivational features than when developing applications that only support users to be more physically active or conduct exercises.

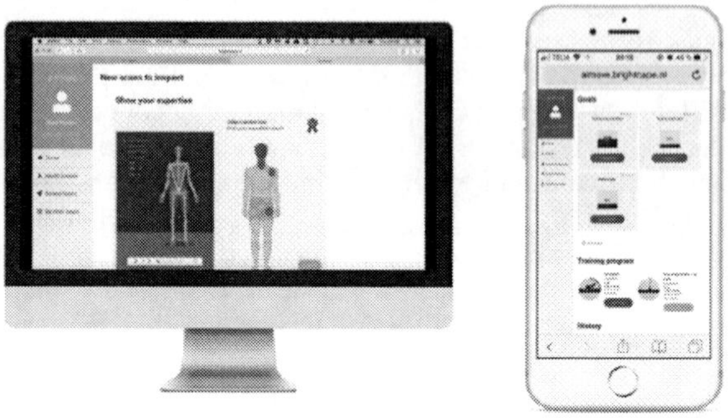

Figure 1. Application for health providers to the left and to their clients to the right.

III. MOTIVATIONAL THEORIES

This section describes central motivational theories and the choices that were made regarding applying some of them in the application developed. Theories not applied in the application are also described below since they are relevant to this domain when the presuppositions for developing are different.

A. *General and Intrinsic Motivational Theories*

Health Belief Model is a psychological model that attempts to provide an explanation of health behavior where the focus is the individual's beliefs and attitudes. It is based on the belief that the individual's perception determines his/her success in conducting a behavioral change [2][3]. In order for the health-promoting behavior to trigger, there must be a stimuli or cue action present. Important factors are: readiness to act, severity, barriers, and self-efficacy.

Self-Determination Theory (SDT) is a general motivational theory. With intrinsic user motivation, it is believed that humans by nature show positive qualities, effort and dedication. People's self-motivation and personal integrity are innate needs (psychological), but they do not happen automatically. There are three needs that have to be met for the intrinsic user motivation to be high: autonomy, competence and psychological relatedness [4].

There are also theories that categorize people in different stages, like *Transtheoretical model (TTM)* and *Precaution Adoption Process Model (PAM)*. These theories can be used to understand the readiness to change of a person. The TTM [5] has six stages of change based on how long the person sustained the new behavior. On the other hand, PAM [6] has seven stages of change based on the psychological state of the person.

B. *Social Motivational Theories*

Bandura's *social cognitive theory* refers to people learning from each other, for example through modelling, observation and imitation [7]. The theory has often been defined as a bridge between behaviorism (Behaviorism is a scientific psychological focus focusing on behavior and learning) and cognitive learning theories since it comprises motivation, memory and attention.

Social comparison theory supports that people, in lack of standard measurements, compare to others for self-evaluation [8], self-enhancement [9], self-projection [10], and coping [11][12]. However, the relation between comparison and competition is rarely studied [13]. Regardless the positive results from psychological studies on social comparison [11][12], often people refuse engaging in comparisons due to social norms [14] or due to different perception of the term "comparison" [15].

C. *Goal setting*

Goal setting has proven to be an effective strategy for encouraging behavioral change [16]. Locke and Latham [17] identify three types of goals: (1) self-set, (2) assigned, and (3) participative set. In terms of self-set goals, the individual is expected to set goals that are realistic to achieve. These type of goals usually have personal significance for the individual since it is related to self-efficacy.

If an assigned goal (a goal set for the individual by someone else) is perceived as motivating, the level of achievement is comparable to a participatory set goal (a goal that the individual has contributed to define). If a goal assigned to the individual does not have a clear motivation, it leads to lower level of achievement.

D. *Applying motivational features in a mobile application for physiotherapy*

In this work motivational features were applied in a mobile application for physiotherapy related exercises. The overall aim with the project and the entire system was to gather, analyze and visualize large amounts of movement related user data. The analyses provided clusters of users with similar movement patterns, where the users could see which cluster they belonged to.

Motivational features were applied based on existing models of motivation. The intrinsic motivational theories described above were difficult to apply within this context since data providing information about internal drivers were not gathered within this framework. Models related to readiness to change could neither be applied in this context. These models demand long-term data related to the user's progress towards an actual behavior change.

Based on the analysis of possible theories to apply, the application developed is focused on features based on social cognitive theory and goal setting with possibilities to set short-term and long-term goals. Features based on social comparison theory were also applied, making it possible to relate own performance with performance of other users. The social comparison was used to see which cluster a user was categorized in, and how many other people that were in the same cluster. Social comparison can also be applied in terms of possibilities for users in the same clusters to exercise together and to support and motivate each other to follow the healthcare professional's exercise recommendations. However, this feature was not implemented in this first version of the application.

As described previously, the features selected and developed were based on practical aspects such as access to data and project duration.

IV. Explorative User Study

As a part of the iterative development, a number of user tests were conducted. One of these tests focused on motivational features and personalization. It was an explorative study where we wanted to get feedback from possible users about how we could integrate motivational features in a meaningful way.

There were seven participants in the test, five men and two women, in an age range between 33 and 52 years. All participants had an education from a university in terms of a Master of Science degree or higher.

The material used were a digital mock-up prototype designed in Figma [18] and a scenario description (Figure 2).

> **Scenario**: You have done the scan and discussed your result with your physiotherapist. Imagine that a hip problem has been detected (or another problem that you want to choose). You have got a training program from the physiotherapist to improve the hip problem and to prevent from hip pain. In the web app you can see what exercises to do and how often as well as the number of repetitions. You can also see the results of the scans.

Figure 2. The scenario presented to the participants.

Semi-structured interviews were conducted, and during the interview questions were asked about motivational features and about personalization. The questions about motivation included questions about what kind of features that would motivate the participants to use the system and what they thought was needed to find motivation to follow the exercise plan. The interview also consisted of questions about getting feedback about progress and about being able to see the progress of other users. Further questions that were given to the participants were questions about personalization and to which extent they wanted the system to be adapted to their preferences and needs. Finally, the participants were asked how they expected the system to support the individualized treatment in physiotherapy. Each interview lasted about an hour. The interviews were recorded, and the data collected was transcribed and thematically analyzed.

A. General motivation

Being able to see progress: The participants described the possibility to be able to see improvement as the most motivating feature, for example by comparing their past scan data with the results from the latest scan or being able to see progress with respect to goals or in terms of reduced pain. Being able to follow the progress was described as one of the most important features, since the lack of progress could be demotivating. In that case, the user could get the feeling of doing something wrong and stop doing the exercises.

Feedback to the user from the physiotherapist: The users had full confidence in the physiotherapist when it came to planning/rehabilitation, but there was a desire to make the planning together with the physiotherapist. The participants thought that frequent interaction with the physiotherapist would increase motivation to continue doing exercises, answer questionnaires and report pain.

Sharing health related information: From a data sharing perspective it was described as important to understand how the system used information provided by the user, for example answers to health related questions. The participants pointed out the importance of a clear connection between questions asked by the system and the feedback that was given. The participants said that if they could not understand this connection, they would hesitate to answer health related questions. Being able to report pain and to get feedback based on pain level was a described as important. This, since one of the main goals for the users is to get rid of the pain.

Reminders: The possibility to get reminders was also described as important among the participants. Especially since it is easy to start to forget doing the exercises when starting to feel better. However, this feature needs to be optional and it has to be possible to enable/disable the reminders.

B. Goal setting

Goal setting was perceived as positive among the participants. However, they were hesitant towards setting their own goals. They perceived the physiotherapists as experts and were expecting them to set the goals.

C. Social comparison

Sharing progress with other users is a feature that some people like and others strongly dislike. For some, it might be too personal to share health related aspects, but for others it is a way of sharing experiences and motivate each other.

In this part, participants were asked to report how it would influence their motivation to see other people's data on their persistence in following the physiotherapist's advice (do the exercises regularly) and in terms of filling in personalized health related questionnaires. Most of the participants (6 out of 7) thought that we asked them to compare their health progress, but it was clarified that we were asking only about their persistence on sticking to the training program or to fill in the health related questionnaires. Their reply was generally that they were uninterested to know about how persistent other users were in following their training programs or filling in their health related questions. However, the participants pointed out that gamification features in the application could make it more interesting to relate to other people's data. An example would be the data used for contributing to a group target or used in competing about being the most persistent user.

Table 1 shows some of the comments the participants shared about their persistence on sticking to the training program or to fill in the health related questionnaires.

TABLE I. COMMENTS FROM THE PARTICIPANTS ABOUT SOCIAL COMPAIRISON

Comments about comparing exercise persistence
A. "If I could see how much I contributed to the group, in a gamified group goal"
B. "So as to get the feeling that you are in this together"
C. "If we collect points together I would be more interested than competing. If other people are persistent then I would be more persistence"
D. "I would be more motivated by competing against the others in the group and try to beat them"
E. "Competition is sometimes good but not here, if you put it closer to collaboration"
F. "It matters more to me if I am doing it than if other people doing it"
Comments about comparing questionnaire completion persistence
G. "If I was the only one did not fill them in, it would have motivated me to fill them in"
Other insights
H. One participant would be compared only to a standard value or a value close to a standard based on a statistical average
I. One participant compared the results of the scan to a colleague to understand how their bodies were crooked, but this was perceived as a comparison promoting awareness, and the participant perceived their reaction to be influenced by the novelty effect with no value to continue comparing future data.

D. *Personalization*

For the participants, personalization was mainly the same as individualized treatment and not related to interaction with application. The interpretation of personalization in this case was an application that generates data to be used to make care related decisions based on the individual's condition and preferences. The users expected the data generation to support the physiotherapist in prescribing the most optimal exercises or treatment for each specific user, and to monitor the rehabilitation progress. In the study, some of the users expected an application like this to enable advanced forms of personalized feedback from the physiotherapist in terms of care progress and potential improvement in condition. Other participants expected that an application like this would generate data in a way that could trigger a personalized intervention based on an input from the user. For example, if the user reports increased pain level, the physiotherapist was expected to react with a personalized intervention in terms of an adjusted treatment plan or with supportive exercise guidance.

V. INSIGHTS AND SUGGESTIONS FOR DEVELOPMENT

Below we present insights from the work in terms of suggestions for development of motivational features for applications in the domain. We also give examples of how we implemented some of the features.

Support communication with the physiotherapist: It is important that the users get individualized feedback based on his/her particular situation. The physiotherapist should monitor progress and make the user aware of that his/her efforts are seen and contribute to the progress. For this application it was considered important to be able to report pain so that the physiotherapist could provide feedback based on the pain level. The implementation of the pain reporting was conducted using a representation of a body and of a pain scale (Figure 3).

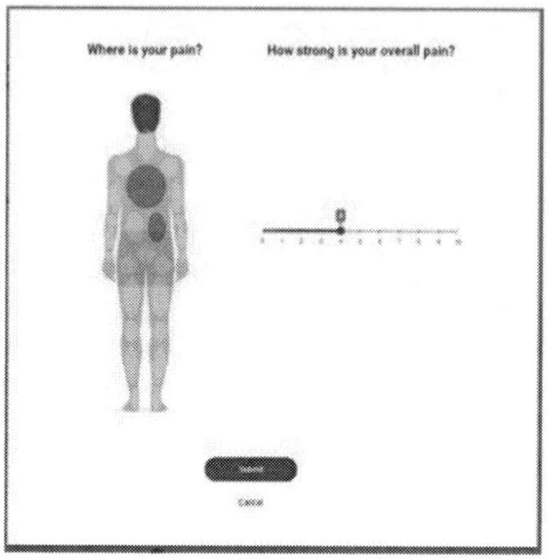

Figure 3. Possibilities to provide the physiotherapist with information about pain level.

Provide feedback about progress: Information about progress is one of the most important features. This can be done by showing improvements in terms of comparing past data with present performance, or in relationship to the goals that have been set. In this application we implemented and visualized the progress of the pain scores (Figure 4). For this user group it was important to follow their progress, and also hopefully be able to see that the pain decreased.

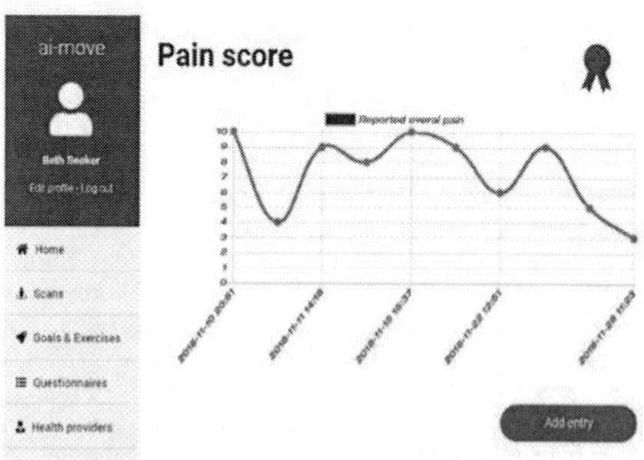

Figure 4. History of the user's pain score.

Support in setting the goals: Goal setting theory has been taken into account in different levels of the application. The goals need to be realistic and have to be based on domain knowledge. Therefore, the goals should be set together with the physiotherapist. Besides being realistic, goals need to be concreate and measurable, and the users could benefit from having explanations to the goals. Finally, it could be motivating to be able to see goals that have been reached. In our application we applied the goal setting features at a high level in terms of self-defined long-term goals (Figure 5), and in terms of possibilities to follow progress in relationship to the goals that have been set (Figure 6). We also applied gamification for short-term goal setting using medals based on the user's compliance in terms of consistency in following the physiotherapist's advices and in doing the exercises (Figure 7).

Figure 5. Possibilities to set own goals.

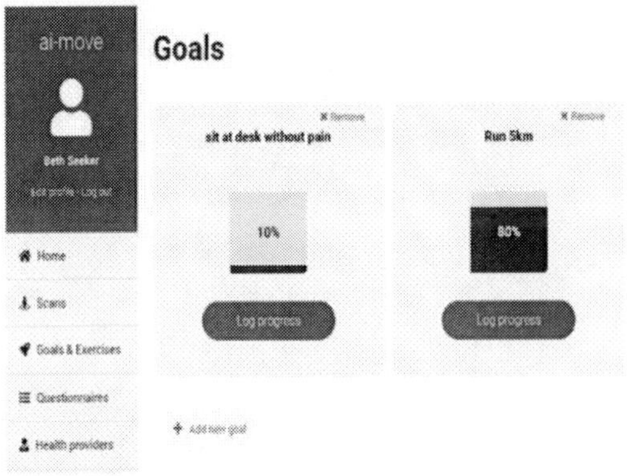

Figure 6. Possibilities to see progress in relationship to the goals that have been set.

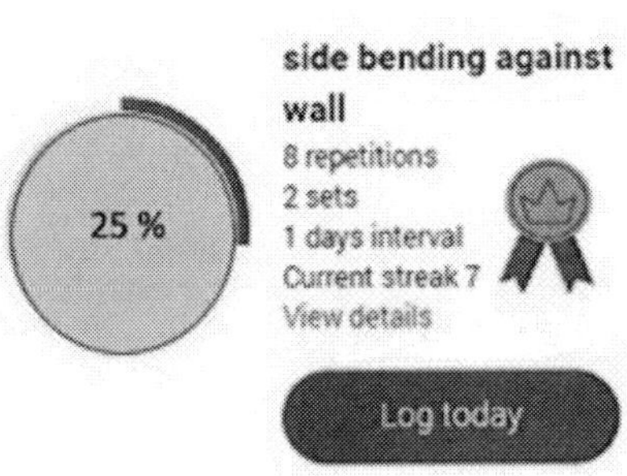

Figure 7. Reward for conduction the exercises every day for a week.

Support from other users: Some users could benefit from sharing experiences with people in the same situation and being able to hear information about other's progress. Users who are unable to keep up with their new exercise routines could benefit from having the possibility to ask for support from people that have managed to engage in new routines [19]. This without triggering competition if that is unwanted. On the other hand, competition might be considered as positive in this case.

Feedback based on data from other users: If the users are categorized into different groups, it could be shown which group the user belongs to and how many people that are in the same cluster. This could provide support in understanding that there are other people struggling with similar issues. Progress could also be based on reported data from other users. For example, it could be shown how successful the suggested exercises are in terms of fast progress.

Comparison in compliance: Comparison can be used for compliance in conducting exercise and in answering health related questions. This can be done regardless of progress and without users sharing sensitive information about their health. For example, rewards for conducting exercises can be given and compared. Another comparable measurement is "number of days after each other that the exercises have been conducted", this shows the user's compliance on a daily basis. Compliance is also a usable motivation aspect when there has been no progress, since it will still be possible to give rewards [20].

Reminders: Motivational messages and reminders can reduce the risk of that exercises will be forgotten when the health improves and the pain has vanished. The motivational messages should be based on the user's actions, for example when reporting that they had conducted an exercise.

VI. CONCLUSION AND FUTURE WORK

Due to the nature of the application, motivational aspects related to goal setting and social motivational theories were the most relevant ones to apply in the development of the application. Goal setting and being able to follow progress were important features to include. It was also shown that the goals for the user's exercises needed to be realistic and set together with the physiotherapist. This was explained in terms of that the physiotherapists were experts in the physiotherapy domain and therefore could estimate true progress. However, it was important that the goals were meaningful and motivating for the user, otherwise it could affect compliance and performance [17].

With respect to being able to compare performance or progress with other users, our results were in line with the research done in the psychological field regarding the rejection of comparison [14][15]. However, if the comparison is disguised as a gamification element, the participants thought that people would be more willing to compare to others for competing, for feeling a part of a group or for contributing to a team. Due to the rejection of comparison in this work, it was impossible to get detailed user specification about design of social comparison features. For example, if they would like to compare to specific individuals, random users of the application or with statistics created by all the users. More research is needed to understand how we can make the users comfortable to talk about comparisons they engage in.

The need for personalization was mainly related to getting qualified feedback from the physiotherapist in terms of him/her following care progress and providing an updated exercise plan. The frequent communication and interaction with the physiotherapist and the individualized exercise plan based on input from the users was described as an import aspect for sharing health related data with the system. The users in our study were willing to provide a variety of personal information, as long as it was used in a meaningful way that supported their progress and their care related decision-making. Other studies have also shown the importance of social interaction and of being seen by the physiotherapist. For some users this social aspect might be the most important motivational feature [20].

Finally, one motivational feature that was not initially discussed with the participants but came up during the interviews was awareness of body posture, and that the visualization of the body in itself could be a motivating feature. This could provide the user with feedback about existing posture and goals showing what to strive for [20].

To summarize, this work conveys insights and suggestions for developing motivational features in applications that supports conducting exercises based on recommendations from a physiotherapist. We have not investigated the use of different motivational theories in a systematic way and do not suggest which motivational theories are most successful to apply in this context. In an exploratory way and for this particular application, practical combinations of different theories were applied. Future work needs to be conducted, both in terms of applying other motivational theories and in terms of evaluating the applied motivational features.

ACKNOWLEDGMENT

The authors would like to thank EIT Digital that funded the project that this work was conducted within. Further, we would like to thank the participants provided us with valuable information during the interviews.

REFERENCES

[1] EUMSUC (2008-2013) Musculoskeletal Health in Europe Report v5.0. EUMSUC, Executive Agency for Health and Consumers. EUMUSC.Net is an information and surveillance network of 22 institutions across 17 countries, supported by the European Community (EC Community Action in the Field of Health 2008-2013) and EULAR.

[2] N. K. Janz and M. H. Becker, "The Health Belief Model: A Decade Later," Health Education Quarterly, vol. 11, no. 1, pp. 1-47, 1984.

[3] M. Becker, "The health belief model and personal health behavior," Slack, vol. 2, no. 4, pp. 324-508, 1974.

[4] S. Asimakopoulos et al., "Motivation and User Engagement in Fitness Tracking: Heuristics for Mobile Healthcare Wearables," Informatics, vol. 4, no. 1, p. 5, 2017.

[5] J. O. Prochaska and W. F. Velicer, "The Transtheoretical Change Model of Health Behavior," American Journal of Health Promotion, vol. 12, no. 1, pp. 38–48, 1997.

[6] N. D. Weinstein, and P. M. Sandman, "The precaution adoption process model and its application," Emerging theories in health promotion practice and research, Jossey-Bass, pp. 16–39, 2002.

[7] A. Bandura, "Health promotion from the perspective of social cognitive theory," Psychology and Health, vol. 13, no. 4, pp. 623–649, 1998.

[8] L. Festinger, "A Theory of Social Comparison Processes," Human Relations, vol. 7, no. 2, pp. 117–140, 1954, doi:10.1177/0018-72675400700202.

[9] R. L. Collins, "For better or worse: The impact of upward social comparison on self-evaluations," Psychological Bulletin, vol. 119, no. 1, pp. 51-69, 1996, doi:10.1037/0033-2909.119.1.51.

[10] J. Suls, R. Martin, and L. Wheeler, "Social comparison: Why, with whom, and with what effect?," Current Directions in Psychological Science, vol. 11, no. 5, pp. 159–163, 2002, doi:10.1111/1467-8721.00191.

[11] S. E. Taylor, J. V. Wood, and R. R. Lichtman, "It Could Be Worse: Selective Evaluation as a Response to Victimization," Journal of Social Issues, vol. 39, no. 2, pp. 19–40, 1983.

[12] J. V. Wood, S. E. Taylor, and R. R. Lichtman, "Social Comparison in Adjustment to Breast Cancer," Journal of Personality and Social Psychology, vol. 49, no. 5, pp. 1169–1183, 1985, doi:10.1037/0022-3514.49.5.1169.

[13] S. M. Garcia, A. Tor, and T. M. Schiff, "The Psychology of Competition: A Social Comparison Perspective," Perspectives on Psychological Science, vol. 8, no. 6, pp. 634–650, 2013, doi:10.1177/1745691613504114.

[14] K. J. Hemphill and D. R Lehman, "Social Comparison and Their Affective Consequences : the Importance of Comparison Dimension and Individual Difference Variables," Journal of Social and Clinical Psychology, vol. 10, no. 4, pp. 372–395, 1991.

[15] F. X. Gibbons and B. P. Buunk, "Individual Differences in Social Comparison: Development of a Scale of Social Comparison Orientation," Journal of Personality and Social Psychology, vol. 76, no. 1, pp. 129–142, 1999, doi: 10.1037/0022-3514.76.1.129.

[16] S. Consolvo, P. Klasnja, D. W. Mcdonald, and J. A. Landay, "Goal-Setting Considerations for Persuasive Technologies that Encourage Physical Activity," Persuasive'09, Article 8, 8pgs, 2009.

[17] E. A. Locke and G. P. Latham, "Building a Practically Useful Theory of Goal Setting and Task Motivation: A 35-Year Odyssey," American Psychologist 57, pp. 705–717, 2002.

[18] "Figma: the collaborative interface design tool," [Online]. Available from: https://www.figma.com/ 2018.12.18.

[19] V. Mylonopoulou, "Design for Health Behavior Change Supportive Technology: Healthcare Professionals' Perspective," NordiCHI '18, pp. 82-92, 2018.

[20] M. Sjölinder et al., "Perspectives on Design of Sensor Based Exergames Targeted Towards Older Adults," In: Zhou J., Salvendy G. (eds) Human Aspects of IT for the Aged Population, Applications in Health, Assistance, and Entertainment, ITAP 2018. Lecture Notes in Computer Science, vol 10927, Springer, Cham, pp. 395-414, 2018, doi:10.1007/978-3-319-92037-5_29.

Enhanced System Usability Scale for Adaptive Courses

Tansu Pancar, Matthias Holthaus, Per Bergamin
UNESCO chair on personalized and adaptive distance education
Swiss Distance University of Applied Sciences (FFHS)
Brig, Switzerland
e-mail: tansu.pancar@ffhs.ch, matthias.holthaus@ffhs.ch, per.bergamin@ffhs.ch

Abstract – **The usage and complexity of learning management systems is increasing continuously. The System Usability Scale (SUS) is a well-known and widely used scale to measure usability of web applications. It is also used in the context of e-learning and several applications were done in popular learning management systems. This study uses a version of the (SUS) translated from English to German on adaptive courses in Moodle and compares the results with the results from a previous study on non-adaptive courses. In addition to comparing results from adaptive and non-adaptive courses, this study also tests the two-factor structure consisting of usability and learnability and suggests two additional items to be included into the learnability factor. In the evaluation of adaptive courses, the total SUS value is lower than non-adaptive courses at our university. In particular, the SUS value for the learnability factor is far lower. It also shows that some items of our German version of the SUS are not suitable for evaluating adaptive courses.**

Keywords-e-learning; usability evaluation; adaptive learning; system usability scale; learnability; moodle.

I. Introduction

Along with the advances in technology, the interaction between education and technology is increasing all over the world. E-learning platforms and applications are being used by many institutions in pure online or blended learning scenarios. Usability of these online tools is an important factor influencing the learning outcome and satisfaction of learners. The System Usability Scale (SUS) [1], with its 10 items, is one of the shortest and most frequently used scales to evaluate web applications. This study uses a version translated into German, which has been validated once before, in our previous study for non-adaptive courses [2].

This current study contains data from university students, participating in adaptive courses in a blended-learning environment in the autumn semester 2018/19. In a first step, the data was analyzed to confirm reliability and validity. In a second step, the results of this study were compared to the results from non-adaptive courses in the previous study. For this, reliability, validity, SUS scores and factor structure were examined and compared. In the original SUS, the two-factor structure contains 8 items for the usability factor and only 2 items for the learnability factor, which does not seem a good distribution. Therefore, the third step was to extend the 10-item scale with two additional questions in order to enhance the learnability factor. Results from the original 10-item scale were compared to the 12-item scale.

Summarizing: the main goal of this study was to measure the usability of adaptive courses. We used the SUS in a two-factor structure with usability and learnability as factors and compared the results of adaptive courses to the results of non-adaptive courses.

Section II presents background information about the SUS from the literature and findings from our previous research on system usability scale with non-adaptive courses. Section III contains a statistical analysis of data and comparison with previous findings. Section III also includes analyses with two additional items for the learnability factor. Section IV discusses findings from both studies. Section V comprises the conclusion and future work possibilities to improve our findings.

II. Background Information and Literature

Usability of web applications and especially learning management systems is a popular research area [3][4]. There are different scales for measuring usability such as the Computer System Usability Questionnaire (CSUQ) [5], the Questionnaire for User Interface Satisfaction, (QUIS) [6] or the Software Usability Measurement Inventory, (SUMI) [7], which contain more questions than the SUS and are more complex to analyze. Translations of the SUS can be found in different languages, including Spanish, French, Greek, and German. In our previous study, we applied and validated a German version of the SUS.

The original SUS is a one-dimensional tool aiming to measure usability. Although not used widely, there are studies that use a two-factor structure, mainly on websites [8][9]. This two-factor structure was tested and validated for the German version of the SUS in an e-learning domain with our previous study [2].

Digital learning systems are considered adaptive if they can dynamically change to better suit the learning in response to information collected during the course of learning rather than on the basis of preexisting information, such as learner's age, gender, or achievement test score, as defined by the U.S Department of Education, Office of Educational Technology [10]. However, digital learning systems can also integrate preexisting information such as gender or the results of entrance tests. It is possible to list three main groups as a basis for adaptation: (1) gender, learning style, or emotions can be grouped as personal characteristics, (2) topics or task difficulty can be divided as content-specific characteristics, and (3) learning time or place can be grouped as context-based characteristics [11]. These groups enable adaptation in different dimensions throughout the learning experience [12].

Task difficulty, actual knowledge prior to taking the course and adaptive support (hints, feedbacks, level of detail) were used in our adaptive courses. The online learning platform used in our study was Moodle, as this is the learning management system used at our university. Adaptive courses in Moodle were different from classical courses as they contained more interactive tasks, specific feedbacks based on learner`s responses and different learning paths for each learner. Depending on the result in a task, a new task was recommended to the learner. The learner was not expected to complete all tasks to be prepared for the exam at the end of the course.

On the other hand, classical courses in Moodle contained the same tasks and learning content for all learners, in the same order, disregarding their previous knowledge level, performance, and progress during the course. Non-adaptive courses proposed the same path for all learners, and learners did not receive any feedback nor recommendations about what to study next.

In our previous study, data from 722 students, enrolled in non-adaptive courses in Moodle, were collected and analyzed in two different semesters (211 students in spring semester 2015 and 511 students in autumn semester 2015/16). Table I summarizes the findings from these two data sets.

TABLE I. FINDINGS FROM PREVIOUS STUDY

N	SUS Score[a]	SD	α	Item Discrimination power	Difficulty
211	62.87	21.74	0.91	0.54 - 0.79	0.58 - 0.67
511	67.51	19.55	0.90	0.51 - 0.79	0.60 - 0.73

Legend: N = number of participants, SD = standard deviation, α = Cronbach's Alpha

a. SUS Score of 10 Items

Exploratory factor analysis of the first data set (n=211) led to a two-factor structure, consistent with previous models [8][9]. The usability factor was formed with the items 1, 2, 3, 5, 6, 7, 8, 9 and the learnability factor with the items 4 and 10. In our previous study, after further confirmatory factor analysis, using Amos [21], of the second data set (n=511) items 5 and 6 were removed due to high residual values. This current study aims to compare results from adaptive courses with the results from the non-adaptive courses.

III. METHODOLOGY AND RESULTS

This section contains statistical analysis of data and compares results from adaptive and non-adaptive courses.

A. Data

The current study contains data from students enrolled in courses on Mathematics, Statistics, and Science. The evaluation of the adaptive courses with the SUS was carried out together with the standard module evaluation that takes place at the end of each course. Module evaluation questionnaire included questions about the course content and the performance of the lecturer. Students were asked to complete the SUS in order to measure the usability, as of the module evaluation questionnaire focused more on the module content rather than the usability of the online-course. To increase completion rate students were asked to take the module evaluation and therefore the SUS during the last classroom unit.

271 students received the survey and 118 of these students completed and returned it, resulting in a high completion rate of 43%. All courses were adaptive and contained pre-knowledge tests and tasks in detailed and non-detailed versions. After each task, the system recommended a new task to the students. All these courses were taught in a blended learning environment, with 80% self-study rate (including online activity using Moodle) and 20% classroom presence with face-to-face interaction with the lecturer and other students.

All online SUS-questionnaire items could be answered on a Likert scale ranging from 1="does not apply" to 5="is absolutely true". Half of the items had an inverted scale. To correctly calculate the SUS value as suggested by Brooke [1] the scale had to be converted from 1-5 to 0-4 and the directionality of the scales had to be consistent. For this, one was subtracted from the value for each odd numbered question, "value - 1", and the value for each even numbered question was subtracted from 5, "5 - value" [1]. The sum of the values for all 10 items was then multiplied by 2.5 resulting in the total SUS score ranging from 0 to 100. The average SUS score for was 55.08 with a standard deviation of 20.20.

B. Analysis of Data

This next part contains the reliability analysis of the scale. Analyses were done using R [13]. The "psych" package [14] was used to check reliability, validity and exploratory factor analysis and the "lavaan" package [15] was used for confirmatory factor analysis.

Cronbach's alpha value is a good measure to estimate internal consistency and therefore reliability of the scale. Cronbach's alpha value was 0.91 for 10 items, which is well above the widely used limit of 0.7 [16]. Cronbach's alpha values for the two factors, 0.91 for usability factor (items 1, 2, 3, 5, 6, 7, 8, 9) and 0.81 for learnability factor (item 4 and 10) were also above 0.7.

Item discriminatory power and item difficulty were used to assert item validity. In order to evaluate item discriminatory power of our scale, we applied the corrected item total correlation test. Corrected item total correlation values were calculated to be between 0.56 and 0.91. These values are acceptable, based on the common assumption of being greater than 0.3 [17]. To calculate item difficulties for each item, we divided the average response for each question with the maximum possible value (which was 5 for our scale). The difficulty of the questions lay between 0.55 and 0.69, which is in the acceptable range of 0.20 to 0.80 [18].

The Kaiser-Meyer-Olkin test was used as a measure of how suitable our data was for factor analysis. A Kaiser-Meyer-Olkin test value of 0.89 was calculated for our study, which is defined as "meritorious" by Kaiser and Rice [19], implying that our survey data was suitable for factor analysis.

Exploratory factor analysis performing maximum-likelihood method supported the two-factor structure with usability and learnability as factors. The usability factor had sums of squared loading of 3.745 and the learnability factor had sums of squared loadings of 2.485. The two factors explained the variance at 62.3%. The results of the exploratory factor analysis signaled a problem in some items (Item 2, Item 6, and Item 8). These items had similar loadings for both factors and could not clearly be classified as belonging to either of the factors. Although the factor structure was supported, the difference between factor loadings were very small for Item 2, Item 6 and Item 8 had 0.071, 0.051 and 0.052 respectively. After checking these items in detail, we found that the wording of the German translation might have been unclear (in the context of adaptive courses) and might have affected students' responses. In the discussion, we will be reflecting on this.

We performed confirmatory factor analysis, keeping this in mind and calculated the values separately for the complete 10 items and the remaining 7 items after excluding items 2, 6, and 8. Removing the three items, increased the Confirmatory Fit Index (CFI) (from 0.896 to 0.966) and the Tucker-Lewis Index (TLI) (from 0.863 to 0.946) and decreased the Root Mean Square Error of Approximation (RMSEA) (from 0.136 to 0.097) and the Standardized Root Mean Square Residual (SRMR) (from 0.073 to 0.05). These values were in compliance with guidelines to have greater CFI and smaller RMSEA and SRMR measures [20]. Removing these items clearly improved the findings from confirmatory factor analysis. Confirmatory factor analysis was also done with a single factor structure to further test if two factor structure performs better, which resulted in a low CFI (0.837) and high RMSEA (0.168) and SRMR (0.09). These values confirmed that using a two-factor structure was suitable for this kind of usability test.

Table II shows the two-factor distribution, with and without items 2, 6 and 8, calculated with Varimax Rotation. When items 2, 6 and 8 were removed, total variance explained decreased to 47.1% composed of 3.043 for usability and 1.666 for learnability (sums of squared loadings). German translation of the SUS items and the original versions in English can also be found in Table II.

In order to check reliability, we recalculated Cronbach's alpha values for all 7 items, 5 usability items and 2 learnability items. These values were found to be 0.86, 0.89 and 0.81 respectively, confirming reliability.

C. *Comparison of Adaptive and Non-adaptive Courses*

In this part, our current results will be compared to the results from the second data set (n=511) of our previous study. As the confirmatory factor analysis was implemented with the second data set only, we used this data set for the comparison.

The overall SUS score for the original 10 items were 55.08 for the adaptive courses (current study) and 67.51 for the non-adaptive courses (previous study).

TABLE II. EXPLORATORY FACTOR ANALYSIS

Item Nr.	Item	10-Item		7-Item	
		F #1	F #2	F #1	F #2
1	Ich würde gerne häufiger Module [a] wie dieses besuchen. I think that I would like to use this system frequently.	**0.78**	-0.09	**0.78**	-0.08
2	Dieses Modul war unnötig kompliziert. I found the system unnecessarily complex	-0.49	0.56		
3	Es war einfach mit diesem Modul zu lernen. I thought the system was easy to use.	**0.78**	-0.29	**0.83**	-0.22
4	Ich brauchte Support fürs Lernen in diesem Modul. I think that I would need the support of a technical person to be able to use this	-0.23	**0.77**	-0.31	**0.65**
5	Die verschiedenen Aktivitäten waren in diesem Modul gut integriert. I found the various functions in this system were well integrated.	**0.84**	-0.21	**0.84**	-0.14
6	Es gab zu viele Ungereimtheiten in diesem Modul. I thought there was too much inconsistency in this system.	-0.42	0.47		
7	Die meisten würden mit diesem Modul sehr schnell zurechtkommen. I would imagine that most people would learn to use this system very	**0.70**	-0.27	**0.70**	-0.26
8	Es war sehr mühsam mit diesem Modul zu lernen. I found the system very cumbersome to use.	**-0.67**	0.62		
9	Ich fühlte mich in diesem Modul sehr sicher. I felt very confident using the system.	**0.64**	-0.36	**0.68**	-0.34
10	Es brauchte viel Vorarbeit, bevor ich mit diesem Modul lernen konnte. I needed to learn a lot of things before I could get going with this system.	-0.08	**0.80**	-0.09	**0.99**

a. Courses are named as "Modules" in our university.

To be able to compare the SUS values with our altered item numbers we standardized the results to be in the same range as the original SUS, namely to be between 0 and 100. Table III below presents these "normalized SUS scores" for learnability (items 4 and 10) and usability (items 1, 3, 5, 7, 9) factors. It is worth noting that, in the previous study, item 5 was removed from the model together with item 6 due to high residuals, but in order to make comparison easier we included item 5 in the Table III results. We observe that, scores for new data set, which is for adaptive courses are lower than that of the previous data set for non-adaptive courses for both usability and learnability related items with similar standard deviations.

TABLE III. FACTOR BASED SUS SCORES

Items for Calculation	Old Data (N=511)		New Data (N=118)	
	SUS	SD	SUS	SD
2 Items (Learnability)	62.13	27.01	48.41	25.92
5 Items (Usability)	68.47	20.00	57.80	21.67

For the confirmatory factor analysis, we compared results from our current study (n=118) to our previous study (n=511). Our previous study reported slightly higher CFI (0.94 versus 0.966) and lower RMSEA (0.070 versus 0.097) values after excluding the low-loading items. Table IV summarizes these findings from confirmatory factor analysis.

TABLE IV. COMPARISON OF CONFIRMATORY FACTOR ANALYSIS

Data Set	Confirmatory Factor Analysis Results			
	N	*Remarks*	*CFI*	*RMSEA*
1	511	8 Items (5 and 6 removed)	0.94	0.070
2	118	9 Items (2, 6 and 8 removed)	0.966	0.097

D. *SUS with 12-Items*

The factor learnability has only two items in the original version of the SUS. In order to strengthen this factor and for a better explanation in the variance, we formulated two additional questions and added them as listed below. Table V presents the new items in English and German.

TABLE V. NEW ITEMS FOR LEARNABILITY FACTOR

Item	Language	New Item
Item 11	German	Ohne Unterstützung (von Kommilitonen, Dozierende etc.) hätte ich diesen Online-Kurs nicht verstanden.
	English	Without support (from fellow students, lecturers etc.), I would not have understood this online course.
Item 12	German	Ich hatte immer wieder Fragen bzgl. dieses Online-Kurses.
	English	I frequently had questions about this online course.

Although there are several definitions of learnability, the improvement in performance after repeated trials can be taken as a simple definition. A good learnability should lead to a high level of proficiency of the user, within a short time and with minimal effort [22]. The original SUS scale has only two items to measure learnability, item 4 states support from technical staff needed and item 10 addresses one`s own effort to be able to use the system.

Newly proposed item 11 "*Without support (from fellow students, lecturers etc.) I would not have understood this online course*", points to support from other users including fellow students and lecturers, but not technical experts, which implies extensive usage and increasing time spent with the system enables enough knowledge to support others.

Item 12 "*I frequently had questions about this online course*" focuses on learner`s reflection about the system and having questions regarding the system as a dimension of learnability.

As a first step of analysis, the reliability of the survey with the added questions was tested. Cronbach`s alpha value was calculated for all 12 items, the remaining 9 items after removing items 2, 6 and 8, the (remaining) 5 items for the usability factor and the 4 items for the learnability factor. Table VI shows these values in addition to item discriminatory power and item difficulty. We applied the same calculation as in part "C" to calculate the "normalized SUS score" for the learnability factor with 4 items (44.76±25.04), which is slightly lower than the value presented in Table III (48.41±25.92) for the "normalized SUS scores" of usability and learnability factors for 511 and 118 participants.

In a next step, exploratory and confirmatory factor analyses including the new items were conducted. All items showed clear distributions in factor loadings (Table VII) and appropriate CFI (0.964), RMSEA (0.084) and SRMR (0.05) values.

TABLE VI. COMPARISON OF QUALITY CRITERIA

Included Items	α	Item discriminatory power	Difficulty
12 Items	0.92	0.59 - 0.9	0.52 - 0.69
9 Items (2,6, 8 excluded)	0.89	0.58 - 0.79	0.52 - 0.69
5 Items (Usability)	0.89	0.74 - 0.83	0.65 - 0.69
4 Items (Learnability)	0.87	0.69 - 0.89	0.52 - 0.62

The usability factor had sums of squared loadings of 3.116 and the learnability factor 2.636, and both factors explained 57.5% of the variance. Table VII presents the factor loadings with the remaining 9 items after removing items 2, 6 and 8 and adding items 11 and 12. The usability factor contains items 1, 3, 5, 7 and 9 and the learnability factor contains items 4, 10, 11 and 12.

TABLE VII. FACTOR LOADINGS AFTER REMOVAL OF ITEMS

Item	Usability	Learnability
1	**0.782**	-0.107
3	**0.798**	-0.312
4	-0.207	**0.913**
5	**0.816**	-0.216
7	**0.724**	-0.19
9	**0.678**	-0.276
10	-0.138	**0.698**
11	-0.261	**0.731**
12	-0.299	**0.717**

Results presented in the tables above, show that the addition of two new questions for the learnability factor increased overall performance of the SUS tool by increasing the total variance explained to 57.5% (31.1% from usability and 26.3% from learnability) from 47.1% (30.4% from usability and 16.6% from learnability). In addition, the RMSEA value was reduced from 0.097 to 0.084, causing almost no change in the CFI (0.966 to 0.964) and SRMR (0.05 for both cases) values.

IV. Discussion of the Results

This study uses a German version of the System Usability Scale, applied to an e-learning content with adaptive courses and compares the results to previous findings for non-adaptive courses. The reliability and validity of the German version of the SUS are intact for both the current and the previous study. This also holds true when two newly created items are added to the scale in order to counterbalance the two-factor structure.

The application of a short questionnaire like the SUS in order to measure the usability of a learning portal has the potential shortcoming of not differentiating clearly between the usability and the content. It is hard for the students to answer some of the questions considering only the usability without including their experiences and feelings about the learning content. In order to make content and usability more distinguishable, a detailed module evaluation was directed to the students prior to the SUS. This module-evaluation contained questions about lecturer, course content, course literature, and classroom presence.

Exploratory factor analysis supported the two-factor structure of the SUS. This is in accordance with our previous study and further literature [6][7]. The exploratory factor analysis further uncovered 3 items with unclear factor loading, items 2, 6, and 8. This lead to an unacceptably high RMSEA value. After exclusion of these three items, the RMSEA was found to be in an acceptable range. The wording of these three items might have been unclear and might have led to some confusion in the assessment of the adaptive courses. For example Item 8: I found the system very cumbersome to use (in German "Es war sehr mühsam mit diesem Modul zu lernen"). We wanted to measure the usability of the module, but in this question, we did not clearly separate the learning content (e.g. the adaptive online tasks) of the module from the usability of the module. Despite answering the detailed module evaluation prior to the SUS questionnaire, students still might have assessed the complicated usability, the more complex online tasks, or even the complexity of the content. This could be a reason for which the factor load for this item is unclear. This supports the unsuitability of this item in our context. The same problem can also be seen in items 2 and 6.

To balance out the two-factor structure of the SUS we added 2 items meant to load onto the learnability factor. The new 9-item SUS (excluding items 2, 6, 8 and including items 11 and 12) had a clear distribution of factor loadings and CFI, RMSEA and SRMR very similar to the 7 item SUS (excluding items 2, 6, and 8) but had a higher explained variance (57.5% instead of 47.1%). These results validate the inclusion of the new items.

It is to be noted, that the total SUS value for adaptive courses was lower than the total SUS value for non-adaptive courses. This might be due to the richness or the voluntariness of the course. The adaptive courses include a variety of learning content, such as the previous knowledge tests, adaptive online tasks and various recommendations. Breaking down the total SUS value both factors were lower for the adaptive courses than for the one on non-adaptive ones. The difference in learnability factor is much bigger, which might mean, that students need more time to find their way around the course. In addition, students are free to use the adaptive learning system and following the recommendations. They are free to leave their adaptive learning path at any time. They do not have to follow the recommendations. All tasks can be accessed without limitations. Consideration of all these options can lead to an additional cognitive load for some students and therefore lead to a decrease in the usability of the course. We are conscious of this problem in usability and are constantly searching for possibilities to improve the course structure and try to present a simpler overview for adaptive courses. As part of this work (for example after changes in the course structure), we need a short questionnaire that students can answer to evaluate if and how the changes affected them. For this, we need the SUS with 12 items.

As the adaptive courses are quite complex a two factor (Usability and Learnability) structure for the SUS made sense. As with the total SUS value both factors were lower in the analysis of the adaptive courses than the one on non-adaptive ones. The difference in learnability factor is much bigger, which might mean, that students need more time to find their way around the course.

From the course-development side, we understand that the first orientation for students in adaptive courses requires more time. Students are accustomed to non-adaptive courses at the university and therefore, they have to adapt to the new adaptive course structure and hence sometimes have questions about how to work with the adaptive courses. We are working on making this transition seamless and rapid and on making adaptive courses more self-explanatory.

User experience was not measured in this current study (2018/19), nor was it measured in the reference study (2015/16). Therefore the user experience could not be compared. Future studies can take user experience into consideration, specially looking at improvements in user experience across studies.

German translation of some items (items 2, 6, and 8) were not clear and in the context of adaptive learning, could cause misunderstanding and misinterpretation. This problem in translation is an important limitation of this study, which should be considered in future studies.

V. Conclusion and Future Works

The blended learning offer of the examined university is based on a concept with around 20% interaction with a teacher (i.e in classrooms) and 80% self-learning time, which is guided and supported by the online learning system Moodle. The introduction of adaptive online courses was primarily designed to support the self-study phase of the students. In this context, the efficient use of the functionalities of the learning platform plays an important role with regard to a high degree of usability and learnability. This is especially important in order to prevent the students from a high cognitive load due to a difficult handling of the learning platform or within the courses, which would affect negatively the learning performance of the students. In fact, the SUS value (55.08) was considerably lower than that of the non-adaptive courses examined (62.87 and 67.51, respectively). This is a distinct indication that the usability of adaptive courses should be increased. If we consider the SUS value of usability (57.8) and learnability (48.41 respectively 44.76) we see, that the value for learnability is much smaller. It takes a longer time for a student to understand the orientation and flow of the course in their first appearance in an adaptive course. We are currently working on improving the clarity of the adaptive online courses and we have started to produce short explanatory videos, explaining how to learn with adaptive courses. We aim to make the process of introduction (learnability) to adaptive course structure easier with these explanatory videos.

The limits of the short scale were also shown here. Accordingly, after calculating the SUS values, we propose to examine more closely how the students have used the course and what progress they have made. We suggest that we have already done this in previous studies, to analyze the log files from the Learning Management Systems database more closely and to relate them to indicators of the students' learning progress. This form of analysis has the advantage that objective data was collected without interfering with the students' learning process, for example by asking questions. This type of data collection also made it possible to efficiently evaluate large datasets or datasets over a longer period of time. In individual cases, these forms of data collection could be supplemented by questions or interviews. To sum up, the SUS has proven to be a reliable instrument also for adaptive courses. However, it also turned out that its German version still has some weaknesses, as demonstrated by the linguistic problems and the exclusion of items. The next step is to test the original scale again in larger samples and test it for potential effecting variables, such as experience with Moodle or satisfaction with IT support or classroom sessions. In our previous study [2], we found that satisfaction with classroom sessions or the lecturer has an influence on the SUS value. These factors may also be important in adaptive online courses.

REFERENCES

[1] J. Brooke, SUS: a "quick and dirty" usability scale. In P. W. Jordan, B. Thomas, B. A. Weerdmeester & A. L. McClelland (eds.), Usability evaluation in industry (pp. 189-194). London: Taylor and Francis, 1996.

[2] P. Bergamin, E. Werlen, M. Holthaus, and M. Garbely, System Usability für E-Learning-Anwendungen "Validierung einer deutschsprachigen Version einer Kurzskala zur Messung der Usability in einem Blended Learning Szenario", unpublished.

[3] R. P. Bringula, "Influence of faculty-and web portal design-related factors on web portal usability: A hierarchical regression analysis", Computers & Education, 68, pp. 187-198, 2013.

[4] N.J. Navimipour and B. Zareie, "A model for assessing the impact of e-learning systems on employees' satisfaction", Computers in Human Behavior, 53, pp. 475-485, 2015 .

[5] J. R. Lewis, "IBM computer usability satisfaction questionnaires: psychometric evaluation and instructions for use", International Journal of Human-Computer Interaction, 7(1), pp. 57-78, 1995.

[6] J. P. Chin, V. A. Diehl, and K. L. Norman, "Development of an instrument measuring user satisfaction of the humancomputer interface", In Proceedings of the SIGCHI conference on Human factors in computing systems ACM Press, New York, pp. 213–218, 1988.

[7] J. Kirakowski and M. Corbett, "SUMI: The software usability measurement inventory", British journal of educational technology 24, 3, pp. 210–212, 1993.

[8] S. Borsci, S. Federici, and M. Lauriola, "On the dimensionality of the System Usability Scale: a test of alternative measurement models", Cognitive processing, 10(3), pp. 193-197, 2009.

[9] J. R. Lewis and J. Sauro, "The factor structure of the System Usability Scale", In Proceedings of the 1st International Conference on Human Centered Design: Held as Part of HCI International 2009, pp. 94-103. Berlin, Heidelberg: Springer-Verlag, 2009.

[10] U.S. Department of Education, Office of Educational Technology, 2013.

[11] K. Wauters, P. Desmet, and W. van Den Noortgate, Adaptive item-based learning environments based on the item response theory: Possibilities and challenges. In Journal of Computer Assisted Learning, 26(6), pp. 549-562, 2010.

[12] M. Holthaus, F. Hirt, and P. Bergamin, "Simple and Effective: An Adaptive Instructional Design for Mathematics Implemented in a Standard Learning Management System," CHIRA 2018 - 2nd International Conference on ComputerHuman Interaction Research and Applications - Proceedings, pp. 116 - 126, 2018.

[13] R Core Team (2018). R: A language and environment for statistical computing. R Foundation for Statistical Computing, Vienna, Austria, https://www.R-project.org/ [Accessed: 01-Oct-2019]

[14] Revelle, W. (2018) psych: Procedures for Personality and Psychological Research, Northwestern University, Evanston, Illinois, USA, https://CRAN.R-project.org/package=psych [Accessed: 01-Oct-2019]

[15] Yves Rosseel (2012). lavaan: An R Package for Structural Equation Modeling. Journal of Statistical Software, 48(2), 1-36, http://www.jstatsoft.org/v48/i02/ [Accessed: 01-Oct-2019]

[16] J. C. Nunnally, Psychometric Theory, 2nd ed., McGraw-Hill, New York, NY, 1978.

[17] E. Cristobal, C. Flavian, and M. Guinaliu, "Perceived e-service quality (PeSQ):measurement validation and effects on consumer satisfaction and web siteloyalty", Managing Service Quality 17 (3), pp. 317–340. 2007.

[18] H. D. Mummendey and I. Grau, "The questionnaire method: basics and application in personality, attitude and selfconcept research", Hogrefe Publishers, 2014.

[19] H. F. Kaiser and J. Rice, "Little Jiffy, Mark Iv". Educational and Psychological Measurement, 34, pp. 111-117, 1974.

[20] D. Hooper, J, Coughlan, and M. Mullen, "Structural Equation Modelling: Guidelines for Determining Model Fit", Electronic Journal of Business Research Methods, 6(1), pp. 53-60, 2008.

[21] J. L. Arbuckle, Amos (Version 23.0) [Computer Program]. Chicago: IBM SPSS, 2014.

[22] M. Unsöld, "Measuring Learnability in Human-Computer Interaction", Masters thesis, Ulm University, 2018.

Are you Lost? Using Facial Recognition to Detect Customer Emotions in Retail Stores

Valter Borges*, Rui P. Duarte*, Carlos A. Cunha* and David Mota†
*School of Management and Technology
Polytechnic Institute of Viseu, Viseu, Portugal
Email: estgv16626@alunos.estgv.ipv.pt, pduarte@estgv.ipv.pt, cacunha@estgv.ipv.pt
†Bizdirect Competence Center, Viseu, Portugal
Email: david.mota@bizdirect.pt

***Abstract*—The understanding of consumer behavior is a dynamic field, especially relevant to the success of companies and for consumer satisfaction. It is especially important in the situation of intense competition, currently characteristic for the retail store industry, where companies fight for every individual customer. Moreover, companies do not want customers to enter their system and leave without buying products they intended to buy. This has an impact on user satisfaction and retail stores income. In this paper we present a method that targets customer satisfaction in the aforementioned context using a facial recognition system acting at the emotional level of the customer. Our method is based on cumulative negative emotions that are associated to a sadness level, which triggers events for retail store assistants to help customers. Results show that this method is adequate to measure these emotions and is a useful reference for retail store assistant intervention.**

Keywords–Image recognition; sentiment analysis; activity recognition; user satisfaction; retail environments.

I. INTRODUCTION

The concept of shopping has been changing during the years. Today shops are not only the place where customers go to buy products but also the place where they spend part of their time. Thereof, retail stores need to adapt to the needs of customers in order to provide them a positive experience. Two perspectives are present: the customer that wants to find and buy a specific product and the retail store that wants to increase sales. Although a relationship can be established between perspectives, in real context scenarios, they have different approaches to achieve a win-win-win solution for the customer–retailer–manufacturer relation. According to Oliver [1], it is more challenging to fidelize an existing customer than to attract new ones. A simple way to lose customers is when they come into a store to buy products they cannot find and leave without buying them. This transforms the process into an unsatisfactory experience for all the players involved, which need to be taken into account.

The application of computer vision techniques in retail dates back to more than two decades [2]. More recently, due to advances in computer vision, machine learning, and data analysis, retail video analytics can provide retailers with much more insightful business intelligence [3][4][5]. Thus it promises much higher business value, far beyond the traditional domain of security, authentication, and loss prevention. Examples include analysis of store traffic, queue data, customer behavior, and purchase decision making, among others. However, it is a real-world scenario, and many technical challenges are present for realistic computer vision techniques: changing and uncontrollable lighting conditions, high-level, complex human and crowd activities, cluttered backgrounds, crowded scenes, occlusion, odd viewing angles, low resolution cameras, limited contrast, and low object discriminability [3].

It is well known that Video Analytics Technology (VAT) mostly focuses on automatic customer detection. However, customer perspective is of most importance since they consume products available in retail stores. One of the potential areas of interest is to determine whether a customer is not finding a specific product. As a consequence, he leaves the store without buying it, which does not relate to a win-win-win situation. If more information about customers is collected using VAT to detect if they are not finding a product and generate triggers to employees informing the occurrence of this, a significant impact in customer satisfaction and retail store sales is foreseen.

Our work focuses on both perspectives and the main contributions are as follows:

- *Emotion Analysis.* To our best knowledge, this is the first scalable attempt to measure negative emotions to determine if a customer is not finding products in a retail store.
- *Real-time notification and intervention.* An integrated platform is developed for the real-time notification of retail stores assistants and intervention with customers.

The remainder of this paper is organized as follows. Section II briefly reviews works in the field of video analytics technology. Section III, identifies the problem to be solved in this work and details our approach, providing a method based on negative emotion analysis. Section IV presents a web interface for the retail store assistant and the method is validated with experimental results carried out in real context scenarios. Section V concludes the paper, providing some hints to future work.

II. RELATED WORK

Our work overlaps with previous research on automatic analysis of human behavior inside retail stores. In this context, several approaches have been studied, as hot zone analysis, automatic activity recognition and sentimental analysis.

The automatic detection of human emotions is a complex problem that has been applied to several ordinary problems. Techniques addressing this problem spans several types of data sources. Faces' images are one of the most promising sources for data analytics related to the emotion detection problem.

A. Hot Zone Analysis

Hot zone analysis aims to identify the trajectory of customers within a store. Trajectory analysis unveils spots with more activity and reveal where customers spend their time. Human's head position estimation was explored to create the initial estimates for tracking algorithms. Zhao et al. [6] presented a method for the detection and tracking of several humans in video frames. They propose boundary and shape analysis for human detection. On top of that, a 3D walking model predicts motion templates from the captured frames to track humans. This work was later improved by Zao and Ram [7], through the inclusion of a detection technique for human identification using Markov chain Monte Carlo methods. The method was tested in indoor and outdoor high-density scenes. In the outdoor scenes, false positives appear at far ends and dense edges. In the indoor scenes, the subtraction method gives erroneous foreground blobs. For human segmentation in both scenes, 1000 iterations are necessary to segment human objects. Leykinv and Mihran [8] developed a method where the human head coordinates are extracted from video frames to determine the position of customers in a store. These coordinates are further used to track customers in video sequences captured in crowded environments. The low-level extraction of the customers in a frame and the use of camera calibration to locate customer's head and location in the picture allows them to infer their location in the store.

B. Activity Recognition

The Activity recognition is related to the shop behaviour and represents the actions of customers when buying products. Monitoring this behaviour is of most importance to academic as to retail stores. Popa et al. [4] analyzed customer behaviour using background subtraction form images. This approach allowed them to detect customers in the entry point and then track them in the system. Popa et al. [5] improved the method for automatic assessment of customer' appreciation of products. First, they classified customer behaviour by participant observation. Next, they implemented the model for motion detection, trajectory analysis, and face localization and tracking for different customers. Sicre and Nicolas [9] resorted to behaviour models for detection of motion, tracking moving objects, and describing local motion. Results have shown that the approach can correctly classify 73% of the frames, for sequences taken in real environments. Later, Frontoni et al. [10] proposed a method to analyze human behavior in shops in order to increase consumer satisfaction and purchases. In their method, they use vertical red, green and blue depth sensors for people counting and shelf interaction analysis. Their results exhibited areas with both positive and negative interactions with products in shelves. They compared their results with ground truth visually recorded and accuracy varies between 97.2% and 98.5%. Hu et al. [11] investigated the detection of semantic human actions in complex scenes. Their work deal with spatial-temporal ambiguities in frames using bag of instances representing the candidate regions of individual actions. A technique based on the combination of Simulated Annealing and Support Vector Machines has shown better results than standard Support Vector Machines.

C. Sentiment Analysis in Videos

Sentiment analysis is another area of video analytics. This type of problems is strongly connected to the problem addressed in this paper, since it determines the emotional level of the customer. Zadeh ett al. [12] addressed this problem using a multimodal dictionary that exploits jointly words and gestures. The approach has shown better results than straightforward visual and verbal analysis. An alternative approach to methods that adopt bag of words representations and average facial expression intensities is presented by Chen et al. [13]. They propose sentiment prediction using a time-dependent recurrent approach that performs fusion of several modalities (e.g., verbal, acoustic and visual) at every time-step. The implementation of the approach using long short-term memory networks has shown significantly improvements over several other approaches. Wang and Li [14] explored sentiment analysis in social media images. The main challenge of the work lies in the semantic gap between visual features and underlying sentiments. Contextual information is proposed to overcome the semantic gap in prediction of image sentiments. The solution was shown effective when evaluated with two large-scale datasets.

III. Approach

Our approach is based on a machine learning system that runs in background for the intervention of retail store assistants with costumers focusing on the analysis of negative emotions using a facial recognition system. When negative emotions are detected, the retail store assistant is notified for customer intervention.

A. Problem Statement

The study of human behavior in retail stores has been carried out in the last years, and their behavior can be interpreted by analyzing their emotional responses [15] to contexts.

At the emotional level, one of the problems that currently exist in customer service is trying to understand their state of mind when inside a store. For that purpose, the detection of emotions from customers will be able to increase the quality of service - the more relevant information about the customer, the better the assistance. The measurement of negative or positive emotions can be carried out by several applications that are available in the market. This work aims at the detection of negative emotions in a time window, where sadness is one of the most significant negative emotion to consider. However, other parameters, like anger, disgust, or fear, are relevant for the measurement of negative emotions. Thus, tracking negative emotions in the context of a store is an open problem, which is of most importance to be solved since it serves the automation of customer-assistant situations, resulting in an increase of the speed of attendance, improve customer satisfaction and increase retail stores sales.

B. Machine Learning Implementation

The performance of machine learning models is deeply dependent on the volume of data available for training models. For that reason, the most accurate models are provided by giants of software that have access to large volumes of data for training models capable of accurate detection of emotions in images. Fortunately, these models are widely available through an Internet accessible API like the IBM Watson [16], Face API [17], Kairos [18], and Amazon Rekognition [19].

In this work, we use Face API [17]. It is a cognitive service developed by Microsoft that supplies algorithms to detect, recognize, and analyze human faces in images. Face API features are obtained in two stages: the first is the detection and recognition of face attributes; in the second stage, a JSON file is returned with the fields that contain face attributes.

The detection stage represents the analysis of the existing faces and returns attributes for each face. When a face is detected, the face rectangle attribute is returned, since it contains the pixels to track the face in the image and gets its bounding box. Within this rectangle, other attributes are returned by the API to the JSON file, namely, face Id, face landmarks, age, emotion, gender, and hair. In this paper, except for face landmarks, all the parameters are considered in two contexts. First, for a general characterization of the costumer, age, gender, and hair attributes are used. These attributes will allow the retail store assistant to better identify the customer (note that for security policies, the system cannot store the face of the customer). Next, for the emotion analysis (cf., Section III-C), the emotion attribute is considered which contains a set of different emotions. The parameters returned by the Face API are a basis of knowledge for the implementation of the emotion tracking method presented in this paper.

C. Emotion Analysis

As previously referred, there are several parameters associated to emotions that are returned by facial recognition systems, namely anger (A_p), contempt (C_p), disgust (D_p), fear (F_p), happiness (H_p), neutral (N_p), sadness (S_p) and surprise (Su_p). In the scope of this work, we only consider negative emotions (A_p, D_p, F_p and S_p) that affect the costumer interaction with the system.

The basic idea of our method is presented in Figure 1. When a customer arrives at a shelf, Face API captures his emotions, and a sadness level β is set to zero. This factor updates in the presence of negative emotions, and once a threshold is passed ($\beta > 50\%$), the assistant is asked to go to the customer. Negative emotions manifest in several ways, and one of the most critical parameters is the sadness parameter, $S_p \in [0..1]$ (values near 1 correspond to the total manifestation of sadness). Therefore, every time a frame captures a customer with a high value of sadness, it may indicate a potential customer not finding a product. Other parameters like A_p, D_p or F_p are also present in negative emotions, and their contribution is analyzed.

To determine the weights to consider in each of the negative emotions, an empiric study (presented in Table II) was carried out with users that were asked to express several emotions: S_p, N_p, D_p, H_p and simulate the action of looking for a product and not finding it, referred to as *Simulated*. In the emotion tests considering H_p and N_p, these parameters have high values, representative of the tested emotion. In the tests for forced sadness and simulation, S_p has low values in most cases, which is justified by the fact that the sadness emotion can result in false positives. However, in this case, the presence of other negative emotions is visible, with small values of A_p, D_p and

Figure 1. Problem specification for negative emotion analysis.

F_p. Analyzing the impact of these parameters in the emotion is an essential factor to determine how to infer sadness when S_p should be naturally present and is not.

In this context, two types of tests were carried out: first, the evaluation of the impact of each negative emotion and, second, the presence of all negative emotions. In the first test, results obtained ($A_p = 47\%$, $D_p = 16\%$, $F_p = 6\%$ and $S_p = 91\%$), show that negative emotion is present in the tests. However, excluding S_p, the other negative emotions cannot be used individually to complement the sadness test, since they are present in a small number of tests which are not representative of the sample. In the second test, we considered the cumulative presence of all negative emotion parameters ($A_p + D_p + F_p + S_p > tol$) for the same scenario (forced sadness and simulation), as shown in Table I.

TABLE I. TOLERANCE TESTS FOR $A_p + D_p + F_p + S_p > tol$.

	Tolerance (tol)		
	0	0.01	0.02
Negative emotions (%)	97.22%	83.73%	80.16%

Results show that when $tol = 0$, 97.22% of the tests reveal the presence of cumulative negative emotions, which is very representative of the tested scenario. The rate decreases for $tol = 0.01$ and $tol = 0.02$. Therefore, when S_p is not representative in a sadness test, the alternative of considering cumulative negative emotions has success rate of 97.22%. Recall that these criteria are used only to improve the success rate of retail store assistants interventions and are used in two contexts: in the evident presence of sadness (high values of S_p) and in the presence of signs of sadness ($A_p + D_p + F_p + S_p > tol$) for low values of S_p. The resulting method is presented in the algorithm depicted in Figure 2. Let $\mathcal{C} = \{c_j\}$, $j = 1 \dots M$, where M represents the number of customers that are detected in the system and $\mathcal{F} = \{f_i\}$, $i = 1 \dots N$, where N represents the number of frames captured in real-time using the Face API for each customer $c_j \in \mathcal{C}$. The algorithm starts by scanning if a customer is detected by the Face API and its faceId is generated. The sadness level of each customer, β_j, is set to zero and frames are captured while the customer is detected in the system. For every captured frame, the Face API returns

TABLE II. USER TESTING IN REAL SCENARIOS: ACTING NORMAL, SIMULATION, FORCE SADNESS, FORCE ANGER AND FORCE HAPPINESS.

Anger (A_p)	Contempt (C_p)	Disgust (D_p)	Fear (F_p)	Happiness (H_p)	Neutral (N_p)	Sadness (S_p)	Surprise (Su_p)	Testing
0	0.001	0	0	0	0.999	0	0	*acting normal*
0.001	0.001	0	0	0	0.985	0.014	0	*simulate scenario*
0	0.002	0	0	0	0.762	0.235	0	*force sadness*
0.004	0.005	0.005	0	0.001	0.962	0.022	0	*sumulate scenario*
0.005	0.002	0.001	0	0.001	0.731	0.261	0	*force sadness*
0	0.002	0	0	0	0.993	0.005	0	*acting normal*
0	0	0	0	1	0	0	0	*force happiness*
0	0.016	0	0	0	0.811	0.172	0	*force sadness*
0.031	0.001	0	0	0	0.967	0.001	0	*simulate scenario*
0.035	0.001	0	0	0	0.966	0.001	0	*force anger*
0	0	0	0	0	0.977	0.023	0	*simulate scenario*
0	0.001	0	0	0	0.905	0.094	0	*simulate scenario*
0	0	0	0	0	0.958	0.041	0	*force sadness*
0	0.089	0.001	0	0	0.58	0.33	0	*force sadness*
0.001	0.027	0	0	0	0.967	0.004	0	*acting normal*
0	0.152	0	0	0.848	0	0	0	*force happiness*
0.172	0.002	0	0	0	0.823	0.003	0	*force sadness*
0.011	0.006	0	0	0	0.962	0.021	0	*simulate scenario*
0.008	0.37	0	0	0	0.621	0.001	0	*force anger*
0.16	0.043	0.001	0	0.001	0.661	0.134	0	*simulate scenario*
0.001	0.025	0	0	0	0.967	0.007	0	*simulate scenario*
0	0.169	0	0	0.009	0.821	0	0	*force sadness*
0.0058	0.011	0	0	0	0.887	0.043	0	*force sadness*
0	0.004	0	0	0.006	0.987	0.004	0	*acting normal*
0	0.001	0	0	0.958	0.04	0.002	0	*force happiness*
0	0	0	0	0	0.857	0.143	0	*force sadness*
0	0	0	0	0	0.84	0.159	0	*simulate scenario*
0.412	0.042	0.09	0.029	0.006	0.57	0.001	0.363	*force anger*
0.001	0.007	0	0	0.001	0.94	0.051	0	*simulate scenario*
0	0.005	0	0	0.038	0.955	0.001	0	*simulate scenario*
0	0.001	0	0	0	0.958	0.041	0	*force sadness*
0	0.001	0	0	0	0.417	0.582	0	*force sadness*
0	0	0	0	0	0.997	0.002	0	*acting normal*
0	0	0	0	1	0	0	0	*force happiness*
0	0	0	0	0	0.965	0.035	0	*force sadness*
0.053	0.004	0	0	0	0.943	0	0	*simulate scenario*
0.127	0.009	0	0	0	0.864	0	0	*force anger*
0	0.0087	0	0	0.036	0.868	0.002	0.007	*simulate scenario*
0	0.001	0	0.001	0.001	0.956	0.033	0.009	*simulate scenario*
0	0.003	0	0	0	0.679	0.318	0	*force sadness*
0	0	0	0	0	0.887	0.113	0	*force sadness*

Algorithm 1: Emotion-based intervention method

Data: $\mathcal{C}$ ◁ detected customers $\mathcal{C} = \{c_j\}$
Data: $\mathcal{F}$ ◁ API frames $\mathcal{F} = \{f_i\}$
Result: β, $\mathcal{I}$ ◁ β = sadness level, $\mathcal{I}$ = Intervention
begin
  foreach $c_j \in \mathcal{C}$ **do**
    $\beta_j \leftarrow 0.0$ ◁ set emotion value to zero
    $\mathcal{I} \leftarrow false$ ◁ no intervention required
    foreach $f_i \in \mathcal{F}$ **do**
      $A_{p_i} \leftarrow A_p \in f_i$ ◁ get anger from f_i
      $F_{p_i} \leftarrow F_p \in f_i$ ◁ get fear from f_i
      $S_{p_i} \leftarrow S_p \in f_i$ ◁ get sadness from f_i
      $D_{p_i} \leftarrow D_p \in f_i$ ◁ get disgust from f_i
      if $(S_{p_i} > 0.5)$ **then**
        $\beta_j \leftarrow \beta + 0.1$ ◁ update sadness level
      else if $(A_{p_i} + F_{p_i} + S_{p_i} + D_{p_i} > 0)$ **then**
        $\beta_j \leftarrow \beta + 0.05$ ◁ update sadness level
      if $(\beta_j > 0.5)$ **then**
        $\mathcal{I} \leftarrow true$ ◁ intervention required
    end
  end
end

Figure 2. Emotion-based intervention method.

negative emotion values that are stored for processing. Every time the algorithm captures evidence of sadness ($S_{p_i} > 0.5$ or signs of sadness ($A_{p_i} + F_{p_i} + S_{p_i} + D_{p_i} > 0$), the value of β_j is updated in a factor of 0.1 or 0.05, respectively. When the sadness level passes a threshold of 0.5, the assistant is informed that a customer needs intervention.

An important consideration is that our system does not retain personal information of a customer. After detection by a camera, only a faceId is generated to uniquely identify the characteristics of that customer. If he leaves the system, the method still continues to try to track the faceId of the customer for five minutes. After that period, the information of the faceId is removed from the database, but the face attributes are kept. With this, personal information of users is not stored, therefore it does not allow the system to track history of a customer. If that customer is again detected in the system, he will be assigned a new faceId.

IV. EXPERIMENTAL DESIGN AND RESULTS

The algorithm presented in the previous section runs in background and processes information that can be visualized by the retail store assistant in an web application (see Figure 3).

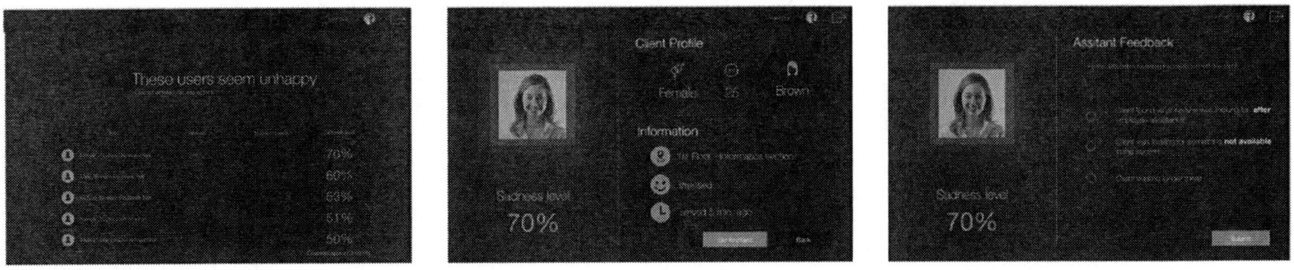

(a) List of customers for assistant intervention. (b) User info and emotions. (c) Assistant feedback.

Figure 3. Web interface for retail store assistant intervention.

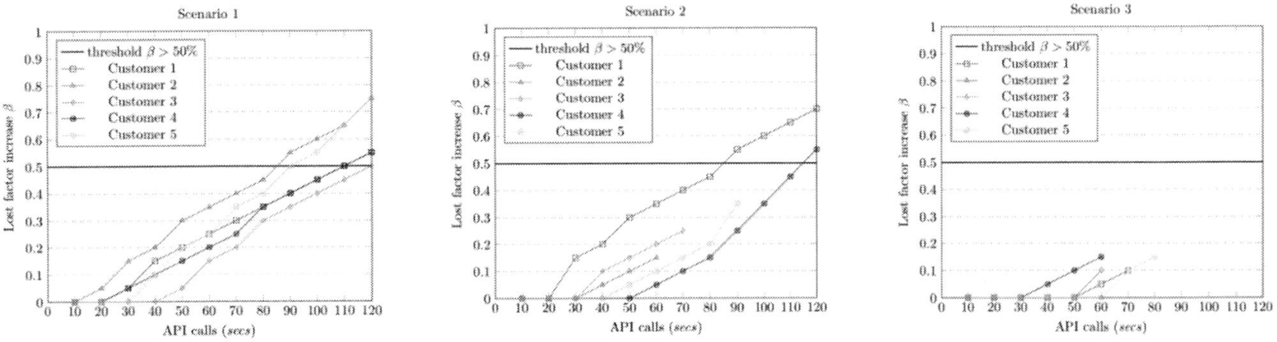

(a) Book is not available in shelves. (b) Book is similar to other products. (c) Book is well identified in shelves.

Figure 4. Results obtained for tests with five customers in the three scenarios.

A. Web Interface

The assistant has access to the notifications management page (see Figure 3a). This page is updated in real time and contains a list of customers that require intervention. Here, some general information of the customers is provided for better identification. When the assistant chooses a customer for intervention, the algorithm running in background stops increasing β for that customer. This value is stored at the moment of intervention, since, otherwise, it would reach 100% for all customers in the time elapsed between the interaction and the time to go to the customers. When the assistant selects a customer, general information is presented (such as hair color, age, gender, location in store, the emotion revealed by the customer and how long the customer is in the system). In addition, the assistant has the possibility of attending the customer and to cancel and return to the call management page as shown in Figure 3b.

The intervention level starts when the assistant clicks in the "go to client" button and the page changes so that feedback data can be provided by the assistant which possesses relevant information regarding the intervention with the customer, as shown in Figure 3c. It is important to note that while the assistant is attending the customer, no further changes in the customer emotions are captured. It is intended to capture the emotions that have caused the customer to exceed the emotional threshold and not to register emotion changes while being under intervention.

B. Results

We have tested the approach described in Section III by carrying out a pilot study. Books were placed in shelves with a camera placed to capture emotions. Five customers where asked to find a book from twenty books in three scenarios:

- Scenario 1: The book is not available in the products placed in the shelves.
- Scenario 2: The book is in the shelves, but very similar to other books, making it difficult to be found.
- Scenario 3: The book is available in the shelves and easy to be identified.

Results obtained are presented in Figure 4. For scenario 1 (Figure 4a), costumers reveal signs of cumulative unhappiness, $(A_{p_i} + F_{p_i} + S_{p_i} + D_{p_i} > 0)$, or sadness $(S_p > 50\%)$ as they realize that they are not finding the product. The sadness level threshold is passed for all customers after a few iterations of API calls. The variation of the sadness level cumulative response is due to the fact that, in the API calls, the customer can reveal one of both negative emotions tested. This implies that there can be an increase of 0.05 or 0.1, depending on the most prevalent negative emotion in each call. In this context, the web interface for the assistant is updated with the data related to the new customer that requires intervention (see Figure 4a). In all cases, the assistant reported option 2 in the feedback page (see Figure 4c). In scenario 2, three customers found the product, after some iterations and left the system.

The other two passed the sadness level threshold. In these cases, the assistant reported option 1 in the feedback page. Finally, in scenario 3, all the customers found the product after a few iterations of API calls, never crossing the sadness level threshold without need for assistant intervention.

To provide flexibility to the system, the assistant can decide the moment of the intervention. As previously referred, when the sadness level threshold is passed, the assistant web page is updated with the customer information. However, if he considers that the sadness level is not increasing with time, he can decide not to go to the customer. If the customer continues to reveal cumulative negative emotions, the assistant then makes the decision to assist him. Moreover, if all assistants are occupied, the system continues to increase the sadness level of a customer, until an assistant is available.

V. Conclusion and Future Work

We presented a novel scalable method based on visual recognition of customer emotions when buying products, using Face API. Our method uses a camera to capture the manifestation of negative emotions at two levels: the effective manifestation of sadness and evidence of sadness, in a set of frames. Our evaluation methodology shows that the method presents good results in real scenarios related to the context of the problem. The implementation of an intuitive web interface allows retail shops assistants to carry out interventions with customers if an emotional threshold occurs. This interface will greatly assist retail stores to have an understanding of which customers require intervention and, in a fast way, provide the necessary help. The natural implications are an increase in sales and customer satisfaction.

Future work will follow two directions. An extension of the method to incorporate face landmarks provided by the Face API will add another layer of decision to our algorithm by tracking the motion of the customer's face to detect if he is not finding a product. Moreover, by introducing Artificial Intelligence (AI), we will be able to anticipate the needs of users based on the previous emotional analysis. Finally, the sadness level was obtained empirically. It will be essential to use AI as a means to adjust this parameter.

Acknowledgment

This work is financed by National Funds through the Research Centre in Education, Technology and Heath (CI&DETS) and the Research Centre in Digital Services (CISeD) of the Polytechnic Institute of Viseu.

References

[1] R. L. Oliver, "Whence consumer loyalty?" *Journal of Marketing*, vol. 63, no. 4, pp. 33–44, 1999, ISSN: 00222429.

[2] R. M. Bolle, J. H. Connell, N. Haas, R. Mohan, and G. Taubin, "Veggievision: A produce recognition system," in *Proceedings Third IEEE Workshop on Applications of Computer Vision (WACV'96)*. IEEE, dec 1996, pp. 244–251, ISBN: 0-8186-7620-5.

[3] J. Connell, Q. Fan, P. Gabbur, N. Haas, S. Pankanti, and H. Trinh, "Retail video analytics: an overview and survey," in *Video Surveillance and Transportation Imaging Applications*, vol. 8663. International Society for Optics and Photonics, 2013, p. 86630X.

[4] M. Popa, L. Rothkrantz, Z. Yang, P. Wiggers, R. Braspenning, and C. Shan, "Analysis of shopping behavior based on surveillance system," in *2010 IEEE International Conference on Systems, Man and Cybernetics*. IEEE, oct 2010, pp. 2512–2519, ISBN: 978-1-4244-6588-0.

[5] M. C. Popa, L. Rothkrantz, C. Shan, T. Gritti, and P. Wiggers, "Semantic assessment of shopping behavior using trajectories, shopping related actions, and context information," *Pattern Recognition Letters*, vol. 34, no. 7, pp. 809–819, may 2013, ISSN: 0167-8655.

[6] T. Zhao, R. Nevatia, and F. Lv, "Segmentation and tracking of multiple humans in complex situations," in *Proceedings of the 2001 IEEE Computer Society Conference on Computer Vision and Pattern Recognition (CVPR 2001)*, vol. 2. IEEE, dec 2001, pp. II–II, ISBN: 0-7695-1272-0.

[7] T. Zhao and R. Nevatia, "Stochastic human segmentation from a static camera," in *Workshop on Motion and Video Computing. Proceedings*. IEEE, Dec 2002, pp. 9–14, ISBN: 0-7695-1860-5.

[8] A. Leykin and M. Tuceryan, "A vision system for automated customer tracking for marketing analysis: Low level feature extraction," in *Human Activity Recognition and Modelling Workshop*, vol. 3. Citeseer, 2005, pp. 6–13.

[9] R. Sicre and H. Nicolas, "Human behaviour analysis and event recognition at a point of sale," in *2010 Fourth Pacific-Rim Symposium on Image and Video Technology*. IEEE, Nov 2010, pp. 127–132, ISBN: 978-1-4244-8890-2.

[10] E. Frontoni, P. Raspa, A. Mancini, P. Zingaretti, and V. Placidi, "Customers' activity recognition in intelligent retail environments," in *New Trends in Image Analysis and Processing (ICIAP 2013)*. Springer Berlin Heidelberg, 2013, pp. 509–516, ISBN: 978-3-642-41190-8.

[11] Y. Hu, L. Cao, F. Lv, S. Yan, Y. Gong, and T. S. Huang, "Action detection in complex scenes with spatial and temporal ambiguities," in *2009 IEEE 12th International Conference on Computer Vision*. IEEE, 2009, pp. 128–135, ISBN: 978-1-4244-4420-5.

[12] A. Zadeh, R. Zellers, E. Pincus, and L.-P. Morency, "Multimodal sentiment intensity analysis in videos: Facial gestures and verbal messages," *IEEE Intelligent Systems*, vol. 31, no. 6, pp. 82–88, Nov 2016, ISSN: 1941-1294.

[13] M. Chen, S. Wang, P. P. Liang, T. Baltrušaitis, A. Zadeh, and L.-P. Morency, "Multimodal sentiment analysis with word-level fusion and reinforcement learning," in *Proceedings of the 19th ACM International Conference on Multimodal Interaction (ICMI '17)*. New York, NY, USA: ACM, 2017, pp. 163–171, ISBN: 978-1-4503-5543-8.

[14] Y. Wang and B. Li, "Sentiment analysis for social media images," in *2015 IEEE International Conference on Data Mining Workshop (ICDMW)*. IEEE, 2015, pp. 1584–1591, ISBN: 978-1-4673-8493-3.

[15] R. Ekman, *What the face reveals: Basic and applied studies of spontaneous expression using the Facial Action Coding System (FACS)*. Oxford University Press, USA, 1997.

[16] "Ibm watson - visual recognition," URL: https://www.ibm.com/watson/services/visual-recognition/ [accessed: 2019-09-29].

[17] "Microsoft cognitive services: Face api," 2019, URL: https://azure.microsoft.com/en-us/services/cognitive-services/face/ [accessed: 2019-09-29].

[18] "Kairos apis and sdks," 2019, URL: https://www.kairos.com/ [accessed: 2019-09-29].

[19] "Amazon rekognition - video and image," URL: https://aws.amazon.com/rekognition [accessed: 2019-09-29].

Assessing Management and Leadership of Work Communities

Related Concepts, Views and ICT-tools

Seppo Väyrynen
Faculty of Technology
University of Oulu
Oulu, Finland
e-mail: seppo.vayrynen@oulu.fi

Heli Kiema-Junes
Faculty of Education
University of Oulu
Oulu, Finland
e-mail: heli.kiema@oulu.fi

***Abstract*— Our key starting point for analyzing and developing work conditions is based on a systematic approach that focuses on humans at work interacting with tasks, organization, technology, and environment, i.e., the work system. This system is not seen in isolation; it is embedded in wider systems of stakeholders, with their contexts enabling or restricting work. Work and wider systems are kept in interactive intra- and inter-connections throughout the key psychosocial and physical human factors within society and work communities. Systems are producing both desirable and undesirable concrete and abstract outcomes. Continuous improvement process aims to achieve optimal work systems to increase desirable outcomes and decrease undesirable ones. Systems, and the inner and outer actors within them, should be optimal and balanced. If this is not the case, productivity and well-being, which include health and safety, are threatened. This paper presents an international collection of guidelines and documentation for promoting and maintaining especially safety (11 maturity models). The documents comprise contemporary, useful, and multi-focused works consisting of clear, important concepts. All the concepts (words) are closely related to safety culture, which in turn is described with both its essential implicit and explicit attributes. The 12th maturity model, the one which the authors have helped develop, is a certain kind of benchmark, as it regards the key aspects of health, safety, environment, and quality (HSEQ). This maturity model is called the HSEQ AP, with AP standing for assessment procedure. In addition to management assessment, this and other models provide good practices and tools for planning, leadership, risk reduction, and more positive features at work, creating a good culture.**

Keywords-safety; health; maturity; management system

I. INTRODUCTION

A team from the University of Oulu has been the main co-developer of the Health, Safety, Environment and Quality Assessment Procedure (HSEQ AP) for supply companies that are partners with big Finnish companies (12 principal companies), particularly those related to the heavy process industry, mainly in the steel, forestry, energy and chemical production sectors, described by Väyrynen et al. [1]. Another key background for this paper is the university's Frictionless Communication (FRICO) project. FRICO aimed at developing managers' and employees' social skills and innovating new actions for organizations to increase work fluency, and overall well-being at work through skillful, professional communication between subordinates and supervisors [2]. This paper, based on a pilot study, aims at analyzing other assessment methods than just HSEQ AP but is based on the key issues of HSEQ AP and FRICO. The paper's one focus is finding ideas to develop the current version of HSEQ AP [3] by emphasizing the role of psychosocial factors (PSFs). We first proposed this PSF improvement of HSEQ AP one year ago, and the key point of that paper was that specific HSEQ knowhow should be distributed to all people at workplaces, not only among the experts, service buyers and various level managers within companies [4]. A wider utilization is considered psychosocially wise, and HSEQ AP-style managerial tools are clearly needed in other manufacturing sectors or service businesses than just the heavy process industry.

Contemporary, more developed managerial and leadership practices and procedures are needed to better meet intra-, and inter-organizational and multi-employer challenges. For example, the PSF and approaches presented in FRICO should be implemented in management systems, with an emphasis similar to Kiema et al.'s [2] and Filppa and Soini's [5] reports, to develop the social skills that are associated with employees' well-being at work and their work engagement. Thus, PSFs should to be more deeply emphasized, generally, e.g., [7][8], and specifically in holistic work systems, cf. Väyrynen and Kiema-Junes [4][6]. The question is what other relevant new features could be found and integrated into HSEQ AP and other management models of good practice. Could the other managerial maturity models, i.e., assessment tools and lists of checkpoints for good HSE(Q) be useful for organizations? Do the other assessment tools include PSFs, and to what extent?

Regarding explicit PSFs, we focus on the proper communication and interaction, individual factors, and subjective well-being. Based on the issues described above, and also considering the methods and results of FRICO, we have determined that the most important words for our analysis are psycho, social, social skills, and communication. Some models that we examined seemed to focus on assessing supply companies, as does HSEQ AP.

We sought answers to the following questions: (1) What key issues do the organizational maturity models typically include? (2) Do these issues cover PSFs that are essential for the whole organization's performance and implementation of the models? (3) Can HSEQ AP be improved through the ideas of the key issues found in the other maturity models and vice versa? (4) How can the assessing methods be further developed by contemporary Information and Communication Technology (ICT) tools, such as smart phones? Regarding the paper's structure: "Section II. Materials and Methods" describes how the simple content analysis of the models, mainly the analysis of the word(s) of interest, was carried out. "Section III. Results" shows the most often and most rarely mentioned words, and then "Section IV. Discussion and Conclusions" discusses the results.

II. MATERIAL AND METHODS

The paper is based on (1) a description of of the documents and field experiences of HSEQ AP and (2) the other managerial maturity model documents we found and analyzed. The other models were sought from various industries, countries, and continents, and they were found by searching a wide range of research- and practice-related publications (cf., Appendix). The key terms were defined by literature review. After finding, choosing, and listing the key words and groups of words, their use frequency in HSEQ AP and the other model documents was determined. Analyzing the documents also included benchmarking. HSEQ AP is related to networking: employees from several supply companies or contractors and self-employed individuals often work simultaneously for the same core production of the principal company, such as in the process industry (the supplier customer). Many of the other models consisted of documents related to the role of networking, i.e., supply companies and contractors. The sub- or support services delivered by supply chain companies typically consist of difficult cleaning services (e.g., cleaning of machinery), maintenance (i.e., repair and service), construction, and security. The principal companies that are buying these services must assure their customers that their facilities satisfy the requirements for holistic quality, e.g., work conditions: negative issues regarding ecology or humans and their community shall be at an acceptable minimum. While regulatory needs must be carefully considered, the need to fulfill ethical and imago aspects are also important. HSEQ AP is an auditing tool for checking the management maturity of the whole network led by the principal company. Hundreds of suppliers have been HSEQ AP audited. The principal companies, and particularly their supply chains, have been able to considerably improve their capabilities and outcomes, e.g., accident situation (Figure 1). Figure 2 shows both the essential principal drivers and assessment categories of HSEQ AP. Based on the safety culture and just culture literature by Dekker [9], and Hudson [10], and Reason [11], as well as our above projects, 11 words were chosen as key terms or concepts for analyzing maturity models. The total number of times these 11 implicitly psychosocial keywords were used within the 11 documents was 1,225 (Table 1).

III. RESULTS

The following four terms, with their respective codes, comprise the most often mentioned word(s), concept(s) among all those selected (Table 1) from the documents (Appendix): IV=report(ing), V= inform(ed, -ing, -ation), VI= learn(ing), train(ing), instruct(ion), X= meet(ing), talk, discuss(ion), involve(ment), participat(e, -ion).

The most often mentioned word(s), listed in the order of the documents from 1 to 11, were:

1) Safety climate and inform(ed, -ing, -ation) and meet(ing), talk, discuss(ion), involve(ment), participat(e, -ion);
2) Safety culture;
3) Report(ing);
4) Learn(ing), train(ing), instruct(ion);
5) Learn(ing), train(ing), instruct(ion);
6) Learn(ing), train(ing), instruct(ion);
7) Repor(ing) and -learn(ing), train(ing), instruct(ion);
8) Repor(ing) and -learn(ing), train(ing), instruct(ion);
9) Repor(ing) and -learn(ing), train(ing), instruct(ion);
10) Change;
11) Learn(ing), train(ing), instruct(ion).

The most rarely mentioned word(s), listed in the order of the documents from 1 to 11, were:

1) Just, fair, and report(ing);
2) Safety climate and satisfactory, -faction and Just, fair;
3) Satisfactory, -faction and just, fair and flexib(ility, -le), resilien(ce, -t);
4) Flexib(ility, -le), resilien(ce, -t);
5) Just, fair and flexib(ility, -le), resilien(ce, -t);
6) Change;
7) Satisfactory, -faction and safety climate and flexib(ility, le), resilien(ce, -t);
8) Just, fair and flexib(ility, -le), resilien(ce, -t) and safety culture;
9) Satisfactory, -faction and inform(ed, -ing, -ation);
10) Safety climate;
11) Just, fair.

The most frequent words presented in the analyzed 11 documents were: reporting and informing, training and learning, and participative human interaction. All the documents had some explicitly psychosocial terms, i.e., the ones including words "psycho, social, or communication". The amount of these words was: (document number in parenthesis): (1) 1, (2) 2, (3) 8, (4) 4, (5) 2, (6) 6, (7) 18, (8)

89, (9) 2, (10) 23, (11) 21. Table 1 and Figure 2 present additional details regarding the results.

IV. DISCUSSION AND CONCLUSIONS

We found many implicitly psychosocial-oriented features in the maturity models. The most frequently used terms matched Reason's [11] recommendations for a better safety culture. Thus, we found significant evidence to recommend utilizing HSEQ AP principles and tools to improve the well-being and culture within work organizations. The FRICO methodology's results in the field and the laboratory are promising. FRICO's Peer Group Counseling was effective for improving social skills in the first trials [2][5], and the new practices were implemented in case organizations in the field [5]. Our results [2][5] also highlighted the importance of social skills in work life and for well-being at work, as many researchers have suggested [7][8]. This finding, related to FRICO, encourages not only including PSFs in models but also utilizing them in implementation and general use within work organizations. This shows a pathway for increasing the role of psychosocially skillful practices for supervisors and subordinates in work communities, as well as all types of employers. The "service and repair" of patients and technological systems for manufacturing consist of similar challenges for appropriate, contemporary management of all employees within their work conditions and communities. These challenges can be successfully met by utilizing the HSEQ AP and FRICO approaches together (e.g., for implementing HSEQ awareness and improvements at all levels of organizations).

Ernst & Young [12], a global consulting company, provided the following view of near-future work changes: "However these models were developed with physical safety in mind at a point in time when we were just being acquainted with the internet, when physical health was prioritized and no-one spoke about mental health." To make the best choices in business, more attention must be paid to holistic maturity models. Da Silva [13] has shown that uncontrolled and unskilled outsourcing in the economy can lead to insufficient risk management in supply companies. The control and prevention of this kind of negative scenario is one of the first priorities. Positive views regarding multi-employer and inter-organizational challenges present, for example, in Finnish health care units [14] can be seen by emphasizing psychosocial approaches, particularly those that were promising for health professionals and patient relationships [2]. The HSEQ AP [1] should be developed further with consideration for psychosocial aspects and for expanding HSEQ AP utilization to all employees [4]. Today, the practice is predominantly the following: Managers, supervisors and experts in the fields of purchasing, as well as health and safety, utilize HSEQ AP information via computer access [4]. The application of HSEQ AP–style possibilities to cases in other sectors, such as health care organizations, should be studied further. Therefore, the present study should be extended to find ways to improve the systems for management and leadership. More detailed benchmarking of the HSEQ AP and the 11 documents would reveal new possibilities for granting every-person ICT access to HSEQ information and for providing ideas for its utilization, for example, by smart phones. Many good recommendations can be found by combining the best practices from the 11 documents. New employee-centric tools, with updated, even real-time HSEQ information in the contexts of daily managerial actions and documented management systems, are increasingly possible and useable with contemporary tailored ICT applications. The HSEQ AP can be briefly outlined as follows: HS issues, 11 assessed categories; e.g., "Managers and supervisors have received occupational safety training targeted to managers, which includes the responsibilities of occupational safety." E issues, nine assessed categories; e.g., "Waste sorting has been instructed and trained and containers for different sections of waste exist. Sorting know-how is included in the orientation." Q issues, 20 assessed categories; e.g., "The company has evidence of systematically developing its own network of suppliers and partners in multiple fields of HSEQ."

The above approaches also open new visions for the leadership of multi-employer, mobile and remote work, and work places. HSEQ information would provide much added value throughout the increased role of PSFs, which are needed to improve the implementation of the systems with a user-centric emphasis and usability goals. Smart workplaces and sites could be enabled through the new increased roles of PSFs and ICT, while still being based on existing management and leadership HSEQ expertise. Furthermore, the 11 concepts found and listed can be called implicitly psychosocial key factors related to HSEQ assessment. If we add the 12th concept, the explicit PSFs we used, i.e., psycho-, social-, and communication-related expressions, we have a dozen of the important concepts (words) regarding the essential features that should be included in an HSE(Q)-style or corresponding assessment of leadership, management, planning, or training in work organizations. Their utilization is mainly based on various ICT support, for example, each employee's personal mobile smartphone [15]16]. ICT may also be used for additional psychosocial purposes, such as praising excellent employees for skilled work [17].

REFERENCES

[1] S. Väyrynen, H. Jounila, J. Latva-Ranta, S. Pikkarainen, and K. von Weissenberg, "HSEQ Assessment Procedure for supplying network: A tool for promoting sustainability and safety culture in SMEs, Chapter 5," in Ergonomics and Human Factors in Safety Management. P. Arezes, and P. V. Rodrigues de Carvalho, Eds. Roca Baton, FL, USA: CRC, Taylor & Francis Group, pp. 83–108, 2016. (cf., www.HSEQ.fi, [retrieved: August, 2019])

[2] H. Kiema, M. Mäenpää, T. Leinonen, and H. Soini, "Peer Group Counseling as a tool for promoting managers'

communication skills in industrial and planning organisations," Proc. The Fourth International Conference on Advanced Collaborative Networks, Systems and Applications (COLLA 2014) IARIA, August 2014, pp. 28-33.

[3] M. Koivupalo, H. Junno, and S. Väyrynen, "Integrated management within a Finnish industrial network: Steel mill case of HSEQ Assessment Procedure," in Integrated Occupational Safety and Health Management – Solutions and Industrial Cases. S. Väyrynen, K. Häkkinen, and T. Niskanen, Eds. Cham, Switzerland: Springer, pp. 41–67, 2015.

[4] S. Väyrynen and H. Kiema-Junes, "Information for managers and experts or communication with all employees within organisations and networks – case HSEQ". Proc. The World Congress on Engineering Asset Management, WCEAM, September 2018. in press. (cf., https://easychair.org/smart-program/WCEAM-2018/2018-09-25.html) . [retrieved: August, 2019]

[5] H. Filppa and H. Soini, Eds. Unpublished FRICO project document. 2014.

[6] S. Väyrynen and H. Kiema-Junes, "Exploring blue- and white-collar employees' well-being at work system: Differences in indicators of physical and psychosocial conditions of occupational groups," International Journal of Sociotechnology and Knowledge Development, vol. 10(2), pp. 14-34, 2018.

[7] W. N. Isaacs, "Dialogic leadership," The Systems Thinker, vol. *10* (1), pp. 1–5. 1999.

[8] A. B. Bakker and D. Xanthopoulou, "Creativity and charisma among female leaders: The role of resources and work engagement". International Journal of Human Resource Management, vol. 24, pp. 2760–2779, 2013.

[9] S. Dekker, Patient Safety: A Human Factors Approach, 1st Edition. Boca Raton, FL, USA: CRC Press, Taylor & Francis Group, 262 p. 2011.

[10] P. Hudson, "Safety Management and Safety Culture The Long, Hard and Winding Road." http://www.caa.lv/upload/userfiles/files/SMS/Read%20first%20quick%20overview/Hudson%20Long%20Hard%20Winding%20Road.pdf, 25 p. 2001. [retrieved: August, 2019]

[11] J. Reason, Managing the risks of organizational accidents. Aldershot: Ashgate, 252 p. 1997.

[12] Ernst & Young, "Why should I trust you? A fresh look at HSE." Ernst & Young AU. https://www.ey.com/Publication/vwLUAssets/EY-why should-i-trust-you-a-fresh-look-at-hse-culture/$FILE/EY-a-fresh-look-at-hse-culture.pdf, 16 p. 2016. [retrieved: January, 2019]

[13] A. J. N. Da Silva, I. M. Almeida, and R. A. G. Vilela, "Outsourcing and Accidents in the Electrical Sector," in S. Trzcieliński, and W. Karwowski, Eds. Proc. The 5th International Conference on Applied Human Factors and Ergonomics AHFE 2014, July 2014, pp. 5832-5844.

[14] T. Pohjosenperä, P. Kekkonen, S. Pekkarinen, and J. Juga, "Service modularity in managing healthcare logistics," The International Journal of Logistics Management, vol. 30(1), pp. 174-194, 2019. https://doi.org/10.1108/IJLM-12-2017-0338.

[15] R. Bye, "Designing mobile user experiences: Disruptive Innovation in railway asset information," in Rail Human Factors: Supporting reliability, safety and cost reduction. N. Dadashi, A. Scott, J.R. Wilson, and A. Mills, Eds. London: CRC, Taylor & Francis Group, pp. 453–460, 2013.

[16] K. Thomas, "Wanted: a WhatsApp alternative for clinicians," British Medical Journal, 360, p. k622, 2018. DOI: 10.1136/bmj.k622. [retrieved: August, 2019]

[17] T. Horishita, K. Yamaura, and M. Kanayama, "Study of effective praise in train driver's workplace," in Rail Human Factors: Supporting reliability, safety and cost reduction. N. Dadashi, A. Scott, J.R. Wilson, and A. Mills, Eds. London: CRC, Taylor & Francis Group, pp. 614–620. 2013.

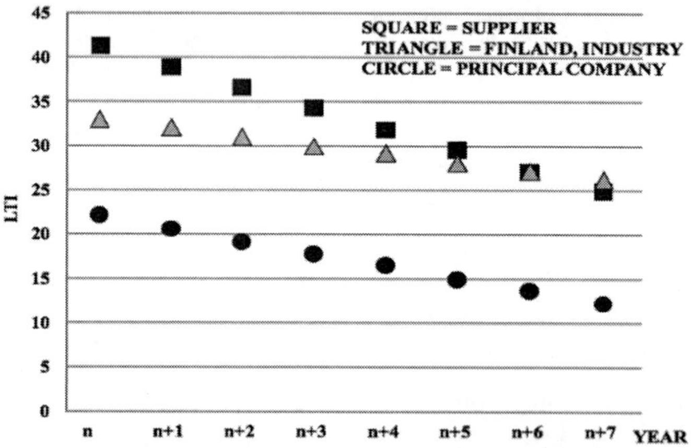

Figure 1. Accident frequency rates (LTI, a linear regression trend) showing an example of the HSEQ AP Cluster's 8-year results (n…n+7). The total rate in the Finnish industry can be seen as well. The positive trend supported and forced with HSEQ AP to supply chain of the principal companies.

Figure 2. The eight drivers (the left list) of the development and utilization of the HSEQ AP (IMS mean Integrated Management System; TQM means Total Quality Management; CSR means Corporate Social Responsibility). The right list describes the nine capability categories chosen for the assessment framework of the HSEQ AP.

TABLE I. THE WORDS CHOSEN TO BE UNDER SPECIFIC CONSIDERATION AS REGARDS ANALYZING AND BENCHMARKING THE MATURITY MODELS.

Code	**Word(s)**	**Amount of mentions**	**Number of maturity documents where used, and an example of use**
I	Satisfact(ory, -tion	51	7, ...product or service quality, and customer satisfaction
II	Safety culture	65	5,...from those with strong safety cultures
III	Safety climate	20	5,... Occupational Safety Climate Questionnaire
IV	Report(ed, -ing)	203	10,...the use of the Global Reporting Initiative
V	Inform(ed,-ing,-ation)	288	11,...Ensuring relevant information
VI	Learn(ing), train(ing), instruct(ion)	61	11, ... Employees have been trained
VII	Flexib(ility, -le), resilien(ce, -t)	11	6,... develops a fair, learning, flexible
VIII	Just, fair	10	7,... must be clear, fair and adhered to
IX	Trust, confident	32	8,... safety management system on the trust
X	Meet(ing), talk(s), discuss(ion), involve(d), participat(ive, -ion)	313	11,... List of toolbox meeting dates
XI	Change	171	8,... being informed of safety-related changes

APPENDIX **No** **Assessment document** **Country(-ies)**	**Aspects:** **H, S, E, Q, Sus-(tainability)** **Scale etc.**	**General (G)** **Specific (S)**	**Number Of pages**	**Reference to document**
1 NOSACQ DK, FI, IS, N, S	S Likert	G	8	http://nfa.dk/da/Vaerktoejer/Sporgeskemaer/Safety-Climate-Questionnaire-NOSACQ50
2 Safety culture maturity model UK	HS Maturity scale Five levels	G	12	HSE (2001). Safety culture maturity model, Health and Safety Executive (HSE), UK.
3 ESPO Green Guide EU	E, Sus Themes, principles	S	38	ESPO (European Sea Ports Organisation) (2012). Green Guide, E, EU, Brussels, BE.
4 RISQS Audit Protocol RSSB Railways UK	H, S, E	S	22	RISQS (2018). Audit Protocol, Industry Minimum Requirements Document no.: RISQS-AP-001 Revision. RISQS Board Industry Minimum, info@rssb.co.uk
5 Score your safety culture Instit. Resilience, sustainability CA, AU	S, Sus (Institutional resilience) -questionnaire, scoring	G, S (transport, aviation)	2	J. Reason (2001). Score Your Safety Culture. *Flight Safety Australia*, January-February, 2001. pp. 40-41. J. Reason (2008). Score Your Safety Culture. TP 13844. (11/2008). *Transport Canada*. 2 p
6 Safety, Health & Environment Checklist for Contractors NL	H, S, E Themes, principles	S (con-tractors, supply chain)	26	SSVV (2018). Safety, Health & Environment Checklist for Contractors, Foundation Cooperation for Safety (SSVV). AK Leidschendam, NL. pp. 27-52.
7 Recommended Practices; S&H Programs in Construction US	H, S Themes, principles	S (con-struction)	40	OSHA (2016). Recommended Practices for Safety & Health Programs in Construction. Occupational Safety and Health Administration. 40 p. www.osha.gov
8 Workplace Safety & Health Manual for Marine Industries SG	H, S Themes, principles, checklist	S (harbours, marine, logistics)	360	WSH Council (2009). Workplace Safety & Health Manual for Marine Industries, SG, 360 p.
9 Vendor Checklist AkzoNobel, H&S, Sustainability, NL	H, S, E, Sus Checklist, Likert-style and yes / no	G (S) Chemical, paints, coatings	4	Akzo Nobel. (2008). Vendor Checklist, Checklist for Supplier Support Visits Concerning Sustainability and HSE issues v3, Sustainability, HSE. Amsterdam, NL.
10 Risk Management Maturity Model (RM3), Road&Rail UK	H, S Maturity scale (five-point)	G, S (transport, road,rail)	64	RM3 The Risk Management Maturity Model, (2017). Version 2.0. Office of Road and Rail, Health & Safety Laboratory, https://www.hsl.gov.uk
11 Niskanen: Research Article: Leadership and OSH processes (risk prevention, collaboration) FI	H, S Data from em-ployers / -ployees	S (chemical industrial manu-facturing)	36	T. Niskanen (2015). Leadership Relationships and Occupational Safety and Health Processes in the Finnish Chemical Industry, In: S. Väyrynen, K. Häkkinen and T. Niskanen (Eds.), *Integrated Safety and Health Management – Solutions and Industrial Cases* (pp. 185-220). Cham: Springer International Publishing.

A User-experience Driven SSIM-Aware Adaptation Approach for DASH Video Streaming

Mustafa Othman
L2TI, Galilee Institute
University Paris 13
Villetaneuse, France
Email: mustafa.othman@univ-paris13.fr

Ken Chen
L2TI, Galilee Institute
University Paris 13
Villetaneuse, France
Email: ken.chen@univ-paris13.fr

Anissa Mokraoui
L2TI, Galilee Institute
University Paris 13
Villetaneuse, France
Email: anissa.mokraoui@univ-paris13.fr

***Abstract*—Dynamic Adaptive Streaming over HTTP (DASH) is a video streaming technique largely used. One key point is the adaptation mechanism which resides at the client's side. This mechanism impacts greatly on the overall Quality of Experience (QoE) of the video streaming. In this paper, we propose a new adaptation algorithm for DASH, namely SSIM Based Adaptation (SBA). This mechanism is user-experience driven: it uses the Structural Similarity Index Measurement (SSIM) as main video perceptual quality indicator; moreover, the adaptation is based on a joint consideration of SSIM indicator and the physical resources (buffer occupancy, bandwidth) in order to minimize the buffer starvation (*rebuffering*) and video quality instability, as well as to maximize the overall video quality (through SSIM). To evaluate the performance of our proposal, we carried out trace-driven emulation with real traffic traces (captured in real mobile network). Comparisons with some representative algorithms (BBA, FESTIVE, OSMF) through major QoE metrics show that our adaptation algorithm SBA achieves an efficient adaptation minimizing both the rebuffering and instability, whereas the displayed video is maintained at a high level of bitrate.**

***Index Terms*—Video Streaming; DASH; ABR; QoE; SSIM; Mobile Networks.**

I. Introduction

Recent studies predicted that the growth of mobile traffic would take up to 20% of total Internet traffic by 2021 [1]. Moreover, it is expected that video traffic will reach 80% of all internet traffic by 2021. The majority of video streaming on the internet today uses the MPEG's Dynamic Adaptive Streaming over HTTP (DASH) standard which aims to deliver video with high Quality of Experience (QoE) [2]–[4]. The principle of DASH consists in dividing the entire video into segments, called *chunk*, in order to send them separately by using HTTP. Each chunk has several versions, each one is encoded with a specific bitrate. Chunks are fetched by clients through HTTP/TCP. This approach makes DASH popular [5] [6] since a) it can be built above the omni-present HTTP and b) client can easily choose, for each chunk, the bitrate which is most suitable to the current network conditions with some Adaptive Bitrate (ABR) algorithm [7]–[11] in order to maximize the user's QoE. Figure 1 depicts the DASH streaming process. Today, major content providers (including NETFLIX and YouTube) use DASH.

Among the main challenges related to DASH scheme [4] [12] [13], there are in particular:

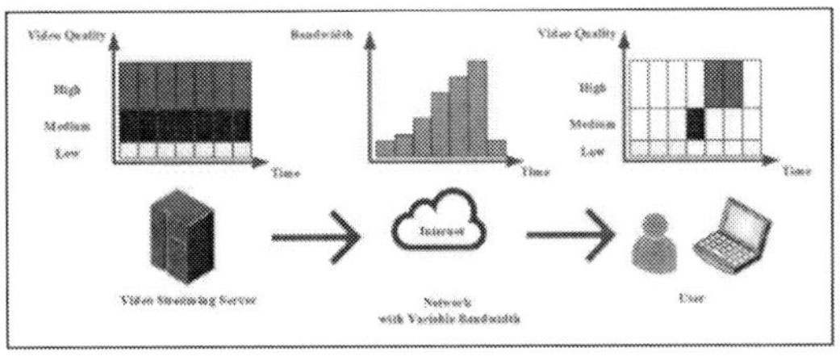

Fig. 1. DASH Streaming Flow Process.

1) *The rebuffering*: this term refers to the freezing of video playback when the buffer is empty and waiting for the next video chunk. The duration and the frequency of rebuffering during a video streaming session is among the most important metrics that affect the user's QoE.
2) *The instability*: it is considered as another important QoE metric. When the bitrate changes from one chunk to the next one, there is surely some variations on the perceived video quality. If the quality of the video among consecutive segments is too abrupt and/or too frequent, the user's QoE will be bad.

It is a real challenge to minimize the rebuffering and instability, and at the same time to maximize the overall quality of the received video. Actually, the surest way to get best stability and avoid rebuffering is to send video always at the lowest bitrate level. However, we would have the worst video quality. So, the selection of the level (in terms of bitrate) of the next chunk is a matter of balance between QoE metrics such as the rebuffering, the instability and the video quality.

Video quality can be assessed through several objective metrics, among which Structural Similarity Index Measurement (SSIM) and Peak Signal to Noise Ratio (PSNR). The PSNR suffers from its inconsistency with the human eye perception video quality. The SSIM is considered to be able to better capture the difference between the original and the encoded images and provides a measurement which is closer to what is visually noticeable as defaults by a human being [11] [14] [15].

This paper proposes a new adaptation algorithm for DASH, namely SSIM Based Adaptation (SBA), by using the SSIM indicator to select the quality of the next video chunk with the objectives of optimizing the QoE by maximizing the video quality and minimizing the rebuffering and instability. To achieve such a balance, the main idea resides in the fact that we decide to a) *increase* the bitrate level only when the SSIM indicates a significant improvement in the video quality (thus getting more video content at almost the same *user perceived* video quality), and b) *decrease* the bitrate level only when there is a real risk of rebuffering (thus minimize the instability).

Our proposal consists in adding the SSIM values for each level of video chunks in Media Presentation Description (MPD), which is the standard way in DASH to provide clients useful informations for video adaptation. Our main contribution is to explore this additional information (SSIM value), in combination with the classical ones, in order to achieve a better adaptation.

Figure 2 helps to better illustrate our basic idea. In this figure, each point gives the bitrate of one of the levels. It can be observed that for chunck number 27, all the levels offer nearly the same SSIM value; whereas for chunck number 140, the lifting in SSIM value for higher levels (hier bitrates) are rather noticeable. This leads to the idea of including the SSIM value among criteria for level selection. Indeed, it is not efficient to select a higher bitrate level when the lower level offers a very comparable video quality.

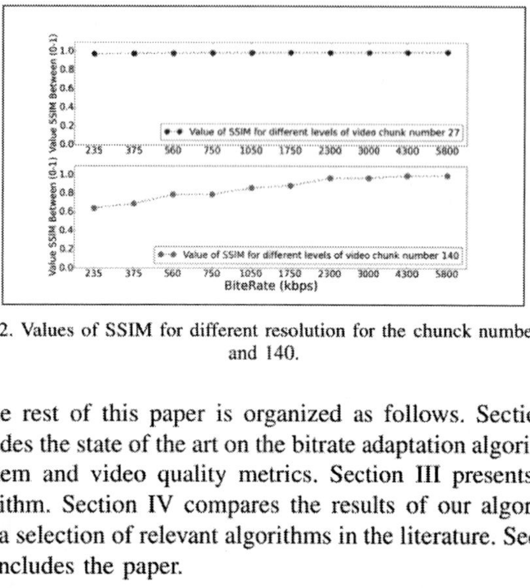

Fig. 2. Values of SSIM for different resolution for the chunck number 27 and 140.

The rest of this paper is organized as follows. Section II provides the state of the art on the bitrate adaptation algorithms problem and video quality metrics. Section III presents our algorithm. Section IV compares the results of our algorithm with a selection of relevant algorithms in the literature. Section V concludes the paper.

II. Related Work

Bitrate adaptation algorithms are generally classified into three categories:

1) *Buffer-based*: the buffer occupancy is used as the main indicator for the selection of the bitrate (level) of the next video chunk to be downloaded [7] [8].
2) *Rate-based*: the bitrate (level) of the next video chunk to be downloaded is chosen to maximize the use of the (estimated) future bandwidth [9] [10].
3) Mixture: a mix of the two previous categories [11].

Huang et al. [7] proposed the Buffer Based Adaptation (BBA) method, which aims to: a) Avoid unnecessary rebuffering and b) Maximize the overall video quality. They used the *buffer occupancy* as a control signal to select the level (bitrate) of the next video chunk instead of the estimated bandwidth. They calculate dynamically a couple of <maximum, minimum> bitrate levels. When possible, the actual video rate keeps growing to the maximum, until there is not enough (with respect to a threshold value) room in buffer, then the video bitrate drops to the minimum level.

Jiang et al. [9] proposed a general bitrate adaptation algorithm framework called Fair, Efficient, and Stable adapTIVE algorithm (FESTIVE). This adaptation algorithm aims to improve fairness, stability and efficiency of the DASH player by predicting throughput to be the harmonic mean of the experienced throughput for the past 5 downloaded chunks, as well as a stability score as a function of the bitrate switches in the past 5 chunks. As the prediction does not take into account the buffer occupancy, FESTIVE may have rebuffering which affects the QoE performance.

The Open Source Media Framework (OSMF) [16] is an HTTP video streaming platform developed by Adobe Systems. OSMF adaptation algorithm selects the next level of video chunk based on the chunk download duration. However, it could lead to instability because of the variability and unpredictability of the bandwidth especially in mobile networks.

Much work has been developed to evaluate the quality of the perceived visual information. Several objective and subjective metrics were proposed with or without reference. Most of them aim to get as close as possible to the human visual system in order to apprehend the different visible distortions (induced by compression, contrast enhancement or transmission...) affecting the original visual information. Among the different metrics, we retain the Structural Similarity Index Measurement (SSIM), considered as an objective metric [17], due to its proven performance. The state of the art shows that this metric has also been selected as a metric to assess the QoE (see e.g., [11] [14] [15] [18]).

In [18], the authors rely on objective QoE metrics as SSIM which is considered as a full reference model. This allows conducting extensive measurement studies and deriving simple relationships applicable to QoE control. They took into account the kind of the videos (e.g., interview, soccer match, movie). Their results showed that they can use SSIM to determine the QoE behaviour of different content, they also shown that video sequences with lower resolution perform better than video sequences with a lower frame-rate.

In [14], the authors incorporated the SSIM into their *Utility Function* as the quality metric. The rational can be summarized as follows. First, the relationship between bitrate and perceptual quality is not linear; as the bitrate increases, the gain in video quality is gradually saturated. Second, the equal division of network bandwidth for video streams of different resolutions (i.e., a vertical line representing a certain bitrate) results in unfair video quality levels as perceived by end-users. In [15], the authors used the SSIM to measure the quality of video transmission through their system (Compressive Distortion Minimizing Rate Control, C-DMRC). The latter uses a distributed cross-layer control algorithm that aims to maximize the received video quality over a multi-hop wireless network with lossy links.

III. Proposed Adaptation Algorithm

This section presents our new adaptation algorithm, referred subsequently as SSIM Based Adaptation (SBA).

A. Rationale

Our algorithm combines the networking level control (buffer-based and rate-based) and the SSIM video quality metric. The key point consists in using the SSIM indicator to determine the level (bitrate and so video quality) of the next chunk to be fetched. A bitrate upgrade takes place not only because it is allowed by the network, but also because it would provide a real gain of *user perceived* quality. The algorithm tries to achieve the following goals :

- By estimating the available bandwidth, we always try to get the best achievable quality.
- We upgrade to a higher bitrate level only when there is a real gain in quality. Thus, we try to maximize the video content downloading (and so minimize rebuffering) at (almost) the same video quality.
- We use buffer-occupancy as rebuffering alert signal and decrease the bitrate, similar to the TCP's behaviour, only when there is a real risk. In this way, we minimize the instability.

The outline of the proposed algorithm is summarized as follows.

- At the beginning of the streaming, as the networking situation is not known yet, the algorithm starts at the lowest bitrate level.
- As the streaming goes on, the algorithm gets a better estimation of the available bandwidth (noted subsequently as EBW).
- The algorithm upgrades to a higher bitrate level not only when it is allowed by EBW, but also because this level provides a real gain on the video quality according to the SSIM indicator.
- In the case where the current EBW is lower than the currently chosen video bitrate level, the algorithm chooses to not decrease the video to a lower bitrate. In this way, the algorithm aims to maximize video quality and to minimize instability due to the bitrate level change. Of course, if the situation persists, there is a risk of buffer starvation.
- The algorithm gives priority to avoiding rebuffering: a critical zone is defined in the buffer. Each time the buffer occupancy falls into this zone, the algorithm decreases the bitrate level to the lowest one.

B. Notation and Conditions

The video is divided into K chunks (video segments) where each chunk has an equal duration of T seconds. The video is encoded at R bitrate levels, denoted as $\mathcal{R} = \{r_j\}_{j=1\ldots R}$ i.e., each chunk has R encoded versions respectively at r_1 to r_R bitrate. By convention, the bitrate levels are ordered as follows $r_1 < r_2 < \cdots < r_R$.

For each chunk and at each bitrate level, the corresponding SSIM metric is computed as follows. The SSIM of each image of the chunk is first computed, then the SSIM of a chunk is deduced as the mean value of the SSIM of the different images of this chunk. The SSIM matrix of the video stream is then computed. It is composed of $Q(i, j)$ element corresponding to the SSIM of the i-th chunk at the r_j bitrate level. From the implementation's viewpoint, this is compatible with the generic DASH framework. Indeed, this SSIM matrix can be pre-computed and stored in the server. It is then sufficient to incorporate it into the MPD (Media Presentation Description) so that the client can get it.

At the client-side, the algorithm estimates the available bandwidth with the following process:

- Each time the client sends the fetch order of a chunk, say chunk i, this time is memorized as s_i. Upon the complete download of the current chunk at t_i with an actual volume of Vb_i bits, the bandwidth actually consumed by the download of chunk i is computed as:

$$CBW(i) = \frac{Vb_i}{t_i - s_i}. \tag{1}$$

- The estimated bandwidth for the fetch of the l-th chunk ($l > 1$), denoted by $EBW(l)$, is then computed as the mean average of the actually consumed bandwidth over the downloaded chunks:

$$EBW(l) = \frac{\sum_{i=1}^{l-1} CBW(i)}{l-1}. \tag{2}$$

For the first chunk, according to the proposed algorithm (as it will be explained later), the lowest bitrate version (r_1) will be used, i.e. $EBW_1 = r_1$.

The algorithm keeps also the trace of the difference in SSIM between the adjacent chunks actually displayed. Let d_i be the level at which the chunk i is displayed. For chunk l ($l > 1$), $\Delta(l) = Q(l, d_l) - Q(l-1, d_{l-1})$, is defined to measure the variation in terms of SSIM related to the previous chunk ($l-1$) when the l-th chunk is displayed. The mean SSIM variation till chunk l (for $l > 1$), denoted by $\alpha(l)$, is then computed as below:

$$\alpha(l) = \frac{\sum_{k=2}^{l} \Delta(k)}{l-1}. \tag{3}$$

Following the convention adopted by the scientific community, the buffer's capacity is given in seconds. The buffer has a capacity of BS (in seconds). It is divided into two regions, region C (for Critical) and region N (for Normal). When the buffer occupancy is below a threshold value (noted by L_c), we are in region C. The current buffer occupancy is denoted by b.

C. Our Algorithm

Hereafter, we describe the SBA algorithm (cf. pseudo-code in Figure 3) which aims to determine the bitrate level (denoted by f) of the next chunk to be fetched.

```
Input:
  R, Q, b, L_c, l, α(l), EBW(l), d_{l-1},
Output: f: Bitrate level at which the next video chunk will
  be fetched
  if b <= L_c then
    f = r_1
  else
    p_l = max{r_j ∈ R, r_j < EBW(l)}.
    if δ(l) = Q(l, p_l) − Q(l − 1, d_{l-1}) > α(l) then
      f = p_l
    else
      f = d_{l-1}
    end if
  end if
  return f ;
```

Fig. 3. Adaptation algorithm.

This algorithm is run each time a fetch order can be issued, i.e. either when a chunk is totally displayed or a chunk is totally downloaded, and of course, when there is room in the buffer (i.e., $BS - b > T$). The parameters $\alpha(i)$ and $EBW(i)$ are assumed to be estimated through parallel processes.

Consider the fetch of the l-th chunk and denote the bitrate level of the previous chunk as d_{l-1} and its SSIM $Q(l-1, d_{l-1})$. This algorithm has two regimes depending on the buffer occupancy:

1) When the current buffer occupancy is in the region C, the next chunk will always be fetched at the lowest level, i.e. $f = r_1$.
2) When the buffer occupancy is in the region N:
   a) With $EBW(l)$, the *potential* bitrate p_l which is the highest bitrate under $EBW(l)$ is determined.
   b) The difference in SSIM if p_l *should be* used, $\delta(l) = Q(l, p_l) - Q(l-1, d_{l-1})$, is then computed.
      i) If $\delta(l) > \alpha(l)$, the algorithm considers that there is sufficient gain in video quality and choose p_l as the bitrate for chunk l, i.e. $f = p_l$.
      ii) Otherwise, the next video chunk at the *current* level (i.e.,. $f = d_{l-1}$) is fetched.

IV. Performance Evaluation

This section compares and discusses the performance of the proposed SBA algorithm to a selection of competitive algorithms.

A. Evaluation Framework

To get a realistic networking context, a set of real traffic situation over the 4G mobile network of a major network provider has been collected. The traces were collected from different areas and periods in Paris to insure a large coverage of the traffic patterns. Three traditional test videos have been used:

1) Animation (Big Buck Bunny) [19],
2) Documentary film (Of Forests and Men) [20],
3) Sport (The World's Best Bouldering in Rocklands, South Africa) [21].

These videos are encoded with FFMPEG codec at the following levels (the ones used by Netflix) [22] [23]: $\mathcal{R}$ = {235kbps, 375kbps, 560kbps, 750kbps, 1050kbps, 1750kbps, 2350kbps, 3000kbps, 4300kbps, 5800kbps}. Videos are then divided into chunks ($T = 4$ seconds, [16]) by using MP4Box-GPAC framework [24].

We developed (in Python) a simulator in order to evaluate the performance of DASH-based adaptation algorithms. This simulator can work in trace-driven mode, i.e., the networking context is reconstituted with real networking traces. The simulator reproduces timely the instants of chunk download completion (which depends on network condition) as well as the chunk playback (which can be blocked by rebuffering). At each instant where the next chunk is to be downloaded, our algorithm enters in action by computing the level of the next chunk.

By using this simulator and the real-traffic trace previously mentioned, we compared SBA algorithm with the following three ones: BBA [7], FESTIVE [9] and OSMF [16]. We have tested two scenarios with two different buffer sizes: a) $BS =$ 120 seconds, b) $BS =$ 240 seconds. Each scenario is tested with 24 different traces. The threshold value (Lc) is set to 12 seconds (3 chunks) in both scenarios.

The performance of the algorithms is assessed through 4 metrics (i.e., Rebuffering, Instability, SSIM, birate). For each metric, the average value is computed on 24 tests:

1) **Average Rebuffering**: is the average of rebuffering (freezing) duration.
2) **Average Instability**: is the average of bitrate changes.
3) **Average of SSIM**: is the average of the SSIM of the video being displayed.
4) **Average of bitrate**: is the average bitrate of the video being displayed.

B. Performance Analysis

This section provides discussions on the achieved performance using the **Animation** video stream. Table I summarizes the results for the two scenarios. For the first 4 lines in the table $BS = 120$ s, whereas for the last 4 lines $BS = 240$ s. One can observe that the SBA algorithm achieves the desired objective with shorter rebuffering, less instability at a good bitrate level.

Moreover, Figure 4 shows that our proposal SBA introduces zero rebuffering for both the scenarios. Actually, we give priority to rebuffering avoidance by setting a critical zone with drastic bitrate drop-off. Being a buffer-based algorithm, BBA works in a similar way and so shows also the same zero rebuffering. On the contrary, FESTIVE and OSMF undergo rebuffering during video chunks playback for 21.208 and 46.25

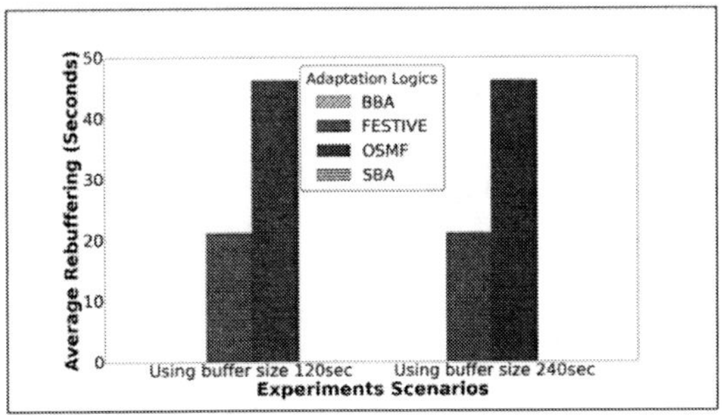

Fig. 4. Average Rebuffering duration for different algorithms with buffer sizes of 120 and 240 seconds and with animation (big buck bunny).

seconds respectively. So, our algorithm performs better than FESTIVE and OSMF for the given scenarios.

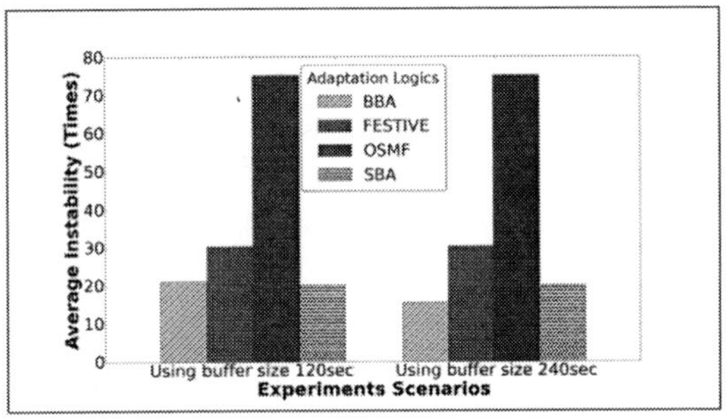

Fig. 5. Average Instability for different algorithms with buffer sizes of 120 and 240 seconds and with animation (big buck bunny).

Figure 5 shows that our proposal SBA achieves good performance, since it is respectively at the first (for $BS = 120$ sec.) and second (for $BS = 240$ sec.) places. For the scenario with $BS = 240$ seconds, BBA algorithm is slightly better than our SBA: this is due to a more conservative bitrate increase approach of BBA. But the price to pay is a much lower average bitrate of BBA compared to the others, where as our algorithm keeps the highest average bitrate (cf. Figure 7).

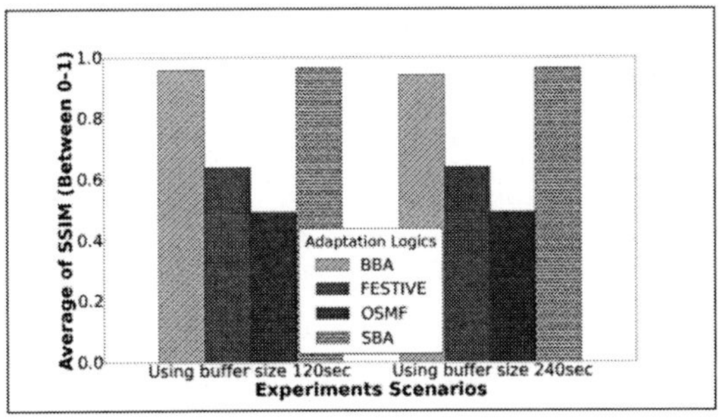

Fig. 6. Average SSIM for different algorithms with buffer sizes of 120 and 240 seconds and with animation (big buck bunny).

As for the SSIM (see Figure 6), our proposal SBA and BBA have similar performance, which is much better than the two others. This means in particular that our choice of upgrading only if there is a real gain in SSIM is justified.

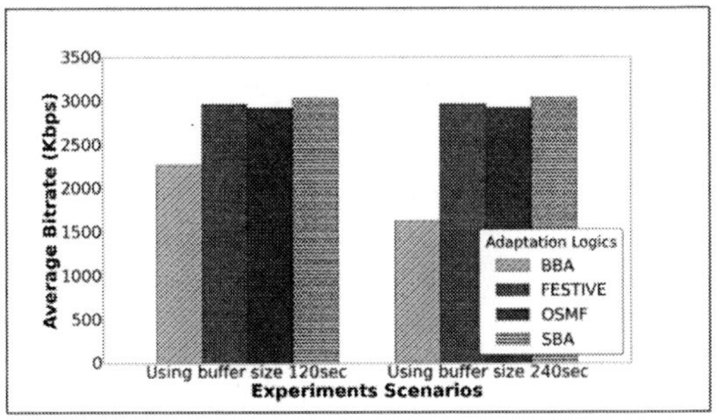

Fig. 7. Average Bitrate for different algorithms with buffer sizes of 120 and 240 seconds and with animation (big buck bunny).

TABLE I. SUMMARIZED RESULTS OF THE TWO SCENARIOS WITH **ANIMATION**

Adaptation Algo.	Rebuffering	Instability	SSIM	BitRate
SBA	**0.0**	**20.25**	**0.967**	**3035.165**
BBA [7]	**0.0**	21.166	0.960	2270.699
FESTIVE [9]	21.208	30.208	0.638	2968.263
OSMF [16]	46.25	75.166	0.492	2926.625
SBA	**0.0**	20.083	**0.964**	**3039.998**
BBA [7]	**0.0**	**15.583**	0.942	1629.180
FESTIVE [9]	21.208	30.208	0.638	2968.263
OSMF [16]	46.25	75.166	0.492	2924.321

TABLE II. SUMMARIZED RESULTS OF THE TWO SCENARIOS WITH **DOCUMENTARY**

Adaptation Algo.	Rebuffering	Instability	SSIM	BitRate
SBA	**0.0**	23.5	**0.941**	2927.646
BBA [7]	**0.0**	**19.916**	0.921	2326.034
FESTIVE [9]	22.625	29.708	0.601	**2963.134**
OSMF [16]	44.625	76.958	0.459	2848.442
SBA	**0.0**	23.416	**0.941**	2929.382
BBA [7]	**0.0**	**14.458**	**0.941**	1635.275
FESTIVE [9]	22.625	29.708	0.601	**2963.134**
OSMF [16]	44.625	76.958	0.459	2841.568

Figure 7 gives the **average bitrate** for different algorithms. As it is shown, our proposal SBA achieves the highest average bitrate for both scenarios. We notices also that BBA, which have similar performance as our algorithm for the first 3 metrics, gets here the lowest bitrate, probably because there is an excessive consideration for rebuffering avoidance.

C. Results Summary

Additional results using the two test videos, namely **Documentary** (see Table II) and **Sport** (see Table III) are given in this section. Similar results to those obtained with **Animation** stream are observed. One can notice that for both scenarios, our proposal SBA algorithm, achieves better ranking for most of the metrics.

V. CONCLUSION

This paper proposed a new adaptation algorithm SSIM Based Adaptation (SBA) for DASH video streaming. This al-

TABLE III. SUMMARIZED RESULTS OF THE TWO SCENARIOS WITH **SPORT**

Adaptation Algo.	Rebuffering	Instability	SSIM	BitRate
SBA	**0.0**	**17.791**	**0.957**	**3086.453**
BBA [7]	**0.0**	17.875	0.952	2470.888
FESTIVE [9]	18.541	28.0	0.673	2992.655
OSMF [16]	46.625	77.083	0.478	2866.815
SBA	**0.0**	17.75	**0.954**	**3078.769**
BBA [7]	**0.0**	**12.708**	0.929	1937.89
FESTIVE [9]	18.541	28.0	0.673	2992.655
OSMF [16]	46.625	77.083	0.478	2864.495

gorithm is user-experience driven since the main control factor is the Structural Similarity Index Measurement (SSIM) which is a good objective indicator for user perceived video quality. This algorithm takes jointly into consideration the networking level indicators (i.e., buffer occupancy, bandwidth) and the SSIM to select the next level of video chunk. The performance analysis of the provided results carried out on trace-driven emulation with real traffic traces (captured in real mobile network) show that the proposed algorithm, compared to some representative algorithms (BBA, FESTIVE, OSMF) through major QoE metrics show that our algorithm (SBA) achieves a more efficient adaptation by minimizing both the rebuffering and instability, whereas the displayed video is maintained at a high level of bitrate. Our main working direction being the joint consideration of networking mechanism and objective video quality metric, we continue in this direction for our future work. For SBA, we plan to explore the impact of the choice of the threshold value of SSIM. We also plan to extend this work to other relevant video quality metrics.

ACKNOWLEDGMENT

This work is supported in part by a scholarship of Campus France (878164H) as well as by a complementary support of the Galileo Graduate School-University Paris 13 (ED146).

REFERENCES

[1] C. V. networking Index, "Forecast and methodology, 2016-2021, white paper," *San Jose, CA, USA*, vol. 1, 2016.

[2] H. Nam, K.-H. Kim, and H. Schulzrinne, "Qoe matters more than qos: Why people stop watching cat videos," in *INFOCOM 2016-The 35th Annual IEEE International Conference on Computer Communications, IEEE*. IEEE, 2016, pp. 1–9.

[3] M. Seufert et al., "A survey on quality of experience of http adaptive streaming," *IEEE Communications Surveys & Tutorials*, vol. 17, no. 1, pp. 469–492, 2015.

[4] F. Dobrian et al., "Understanding the impact of video quality on user engagement," in *ACM SIGCOMM Computer Communication Review*, vol. 41, no. 4. ACM, 2011, pp. 362–373.

[5] F. WANG, Z. FEI, J. WANG, Y. LIU, and Z. WU, "Has qoe prediction based on dynamic video features with data mining in lte network," *Information Sciences*, vol. 60, no. 042404, pp. 1–042 404, 2017.

[6] F. Hartung, S. Kesici, and D. Catrein, "Drm protected dynamic adaptive http streaming," in *Proceedings of the second annual ACM conference on Multimedia systems*. ACM, 2011, pp. 277–282.

[7] T.-Y. Huang, R. Johari, N. McKeown, M. Trunnell, and M. Watson, "A buffer-based approach to rate adaptation: Evidence from a large video streaming service," *ACM SIGCOMM Computer Communication Review*, vol. 44, no. 4, pp. 187–198, 2015.

[8] C. Zhou, C.-W. Lin, X. Zhang, and Z. Guo, "Buffer-based smooth rate adaptation for dynamic http streaming," in *2013 Asia-Pacific Signal and Information Processing Association Annual Summit and Conference*. IEEE, 2013, pp. 1–9.

[9] J. Jiang, V. Sekar, and H. Zhang, "Improving fairness, efficiency, and stability in http-based adaptive video streaming with festive," *IEEE/ACM Transactions on Networking (TON)*, vol. 22, no. 1, pp. 326–340, 2014.

[10] S. Akhshabi, L. Anantakrishnan, A. C. Begen, and C. Dovrolis, "What happens when http adaptive streaming players compete for bandwidth?" in *Proceedings of the 22nd international workshop on Network and Operating System Support for Digital Audio and Video*. ACM, 2012, pp. 9–14.

[11] S. Cicalo, N. Changuel, R. Miller, B. Sayadi, and V. Tralli, "Quality-fair http adaptive streaming over lte network," in *2014 IEEE International Conference on Acoustics, Speech and Signal Processing (ICASSP)*. IEEE, 2014, pp. 714–718.

[12] S. S. Krishnan and R. K. Sitaraman, "Video stream quality impacts viewer behavior: inferring causality using quasi-experimental designs," *IEEE/ACM Transactions on Networking*, vol. 21, no. 6, pp. 2001–2014, 2013.

[13] R. K. Mok, E. W. Chan, X. Luo, and R. K. Chang, "Inferring the qoe of http video streaming from user-viewing activities," in *Proceedings of the first ACM SIGCOMM workshop on Measurements up the stack*. ACM, 2011, pp. 31–36.

[14] P. Georgopoulos, Y. Elkhatib, M. Broadbent, M. Mu, and N. Race, "Towards network-wide qoe fairness using openflow-assisted adaptive video streaming," in *Proceedings of the 2013 ACM SIGCOMM workshop on Future human-centric multimedia networking*. ACM, 2013, pp. 15–20.

[15] S. Pudlewski, A. Prasanna, and T. Melodia, "Compressed-sensing-enabled video streaming for wireless multimedia sensor networks," *IEEE Transactions on Mobile Computing*, no. 6, pp. 1060–1072, 2012.

[16] M. Riad, H. Abu-Zeid, H. S. Hassanein, M. Tayel, and A. A. Taha, "A channel variation-aware algorithm for enhanced video streaming quality," in *Local Computer Networks Conference Workshops (LCN Workshops), 2015 IEEE 40th*. IEEE, 2015, pp. 893–898.

[17] Z. Wang, L. Lu, and A. C. Bovik, "Video quality assessment based on structural distortion measurement," *Signal processing: Image communication*, vol. 19, no. 2, pp. 121–132, 2004.

[18] T. Zinner, O. Hohlfeld, O. Abboud, and T. Hoßfeld, "Impact of frame rate and resolution on objective qoe metrics," in *2010 second international workshop on quality of multimedia experience (QoMEX)*. IEEE, 2010, pp. 29–34.

[19] "Big buck bunny homepage," 2019, URL: https://peach.blender.org/ [retrieved: 2019-05-07].

[20] "Of forests and men," 2019, URL: http://www.offorestsandmen.org/ [retrieved: 2019-05-09].

[21] "The world's best bouldering in rocklands, south africa," 2019, URL: http://www.lt11.com [retrieved: 2019-05-13].

[22] "Ffmpeg homepage," 2019, URL: https://www.ffmpeg.org [retrieved: 2019-06-01].

[23] "Netflix encoding bitrate," 2019, URL: https://medium.com/netflix-techblog/per-title-encode-optimization-7e99442b62a2 [retrieved: 2019-05-23].

[24] "Gpac homepage," 2019, URL: https://gpac.wp.imt.fr [retrieved: 2019-06-05].

Influence of Performance Expectancy, Experience and Perceived Risk on the Usage of Cryptocurrency Investments

Erik Massarczyk, Peter Winzer, Finn Jakob Müller
Faculty of Design – Computer Science – Media
RheinMain University of Applied Sciences
Wiesbaden, Germany
Email: erik.massarczyk@hs-rm.de, peter.winzer@hs-rm.de

Abstract—**The technological development of the blockchain technology allows a new way of processing secured transactions and payments between different parties. Therefore, it is not surprising that new virtual currencies are developed to open new payment methods, as well as investment opportunities. To estimate the intention to buy and use cryptocurrencies, an empirical analysis was performed. The question of research is whether an investment in cryptocurrencies is primarily made for speculative reasons or because of a belief in the establishment of a digital currency. The ordinary regression analyses on base of survey data, which was distributed online, outlines that the intention to use cryptocurrencies is mainly driven by investment purposes.**

Keywords-investments; cryptocurrencies; risk; expierenice; performance expectancy, UTAUT2.

I. INTRODUCTION

Cryptocurrencies have achieved market capitalization of currently around 250 billion euros due to the strong growth in recent years [1]. On the one hand, investors see cryptocurrencies as an opportunity to reach high revenues accompanied with a specific (potentially high) risk, while on the other hand, researchers and experts see cryptocurrencies as opportunity to create a new known and general accepted currency and payment method [2][3]. Therefore, it will be analyzed what most private customers/users think about cryptocurrencies (e.g., Bitcoin or Ethereum) and how cryptocurrencies are used. To estimate the described customer behavior, a quantitative research using an online survey is applied. The resulting database will be analyzed using statistical techniques for data estimation and the statistical program Statistical Package for the Social Sciences (SPSS), which targets on the estimation of results about the later described hypotheses.

In this respect, the paper is structured as follows. In Section II, (a) cryptocurrencies, (b) blockchain, (c) digital versus traditional currencies, (d) challenges, as well as (e) the used research model will be described. Following this section, the methodology and the theoretical approach for carrying out the analysis, will be explained. In Section IV, the results of the empirical analysis are presented. The paper concludes with a summary of the results in Section V.

II. TECHNICAL BACKGROUND

The following background section covers the definition of the research objectives cryptocurrencies and blockchain as well as the used research and conceptual model as well as the challenges in the named research field.

A. Cryptocurrencies

Although the first ideas to develop a digital and anonymous currency date back to 1989 [4], the first virtual cryptocurrency was implemented in 2008 [5], when Nakamoto published an approach for an electronic payment system and a new currency "Bitcoin" based on blockchain technology [6]. This approach differed from earlier approaches in particular in that all transfers must be validated by the community. This validation was performed decentral using a synchronized blockchain across multiple users [7]. To this extent, no third party is required as an intermediary to carry out secured transaction. This means that the currency Bitcoin was created primarily with the intention that transmissions can be cryptographically secured and tracked [5][8]. In addition, cryptocurrencies based on blockchain technology are implemented to (a) guarantee fast worldwide money transfers, (b) establish the privacy of the participating parties through anonymity, and (c) advance the development of a payment system independent of the traditional banking system [3].

B. Blockchain

Following Nakamoto [5], a blockchain is a continuously expandable list of data records, called "blocks", which are linked together by cryptographic methods. Each block typically contains a cryptographically secure hash of the previous block, a timestamp and transaction data [9]-[11]. The blockchain allows the linking of transfers within a decentralized platform, which is distributed and publicly assessable [12][13], where through recording of transfers, processes and information are secured by cryptographic techniques [10][14][15]. The fact that a large number of users of the blockchain can access and track the linked blocks within the blockchain creates confidence in the reliability of the digitally applied processes and transfers [7][10][11][13][15]. Finally, blockchain provides a solution for a trusted, secure, decentralized and (by consensus) peer-validated approach [16]. In general, the entire database is embedded in a peer-to-peer network architecture with equal nodes. Due to the node principle, the system is not dependent on a central location, which could be a single point of failure [3].

Since, the information and data are implemented in the blockchain, which is decentralized distributed, no information can get lost [13]. Any implemented block is irreversibly linked to a previously block and cannot be deleted. Each block contains information about transactions and information of the previous block [5]. A new block is only added in case the verification through the validation and consensus process by the community is done [17]. Any update needs to be performed in a new developed block, which needs to be verified by the described process [18].

The application of blockchains guarantees a technically secure communication on the base of mutual authentication, as well as tamper-resistant asymmetric cryptography, which enables an information exchange by timestamped and logged records [7][12][19][20][21][22]. The blockchain approach implies the irrevocability of changes, i.e., the blocks or information remain permanently in the system and cannot simply be deleted [7][18]. The security mechanisms are implemented to avoid any spam and denial-of-service attacks [23].

The interaction of users within the blockchain takes place by using a related key pair, which comprises a private key and a public key [24]. The latter is publicly visible and comparable to an address that each node has; it can be regenerated for each transaction in order to maintain anonymity. If a node wants to create a transaction and, e.g., add new data, this can be done anytime autonomous by signing it with its (secret) private key [25]. It is then sent to all nodes of the peer-to-peer network to reduce single point of failures [15]. Each node is then able to use the public key to verify the node that created the transaction before a distributed consensus mechanism regulates the addition of the new block [26][27]. A consensus mechanism implemented through the Distributed Ledger Technology ensures that there is only one next block, which is necessary to obtain integrity of the blockchain [15][26][28. This means that the consensus mechanism ensures that the transactions and blocks are sorted chronologically, which verifies the integrity, coherence and consistency of the blocks sustaining in the blockchain [7][15][19].

A subsequent update process ensures that all participants always have the latest version of the database at their disposal [29]. There are several methods for validating the transaction and reach consensus. The most common of which are currently known as 'Proof of Work' and 'Proof of Stake'. In these two methods, hash values are generated by the network nodes according to a certain pattern. Depending on the length of the blockchain, the degree of difficulty and the computing power required for this increase. In this context, the working nodes are also referred to as 'miners' [27]. The type of the utilized consensus mechanism varies in dependency from the type of network and other factors [26]. In summary, when a transaction is validated, it is stored in the block and chained in the blockchain [15], with the community deciding on the validation. I.e., this validation could only be manipulated by someone who has control over the majority (> 50%) of nodes, which is extremely unlikely due to the worldwide decentralized networks [7]. The timestamp documents (transparently for the whole network) the time of implementation and adjustments [30].

C. Digital versus Traditional Currencies

The main differences between traditional and digital currencies are: (a) The digital currencies are organized decentral using block-chain technology and do not require banks or other intermediaries (unlike traditional currencies). (b) The digital currencies are (uniformly) valid and available worldwide [31], while the traditional currencies are generally specific to individual states or economic areas [32]. The use of traditional currencies (especially for international transactions) results in relatively high transaction costs, whereas digital currencies cause no or only very low transaction costs due to the consensus mechanism and the very fast "automatic" validation of transactions [3][31][33]. (c) Digital currencies offer a high degree of anonymity and protection of personal data, which is not provided by traditional currencies (e.g., credit card payments or money transfers). In traditional currencies, this anonymity could only be achieved through cash payments, but the transaction costs are very high. In addition, cash payments are strongly limited or regulated in many countries

Another central feature of a currency is that it is always available, transportable and divisible. This is also true for cryptocurrencies [34].

In contrast to the traditional currencies, each cryptocurrency has a fixed limit regarding the maximum currency units that can be issued [34].

D. Challenges

Due to this "gap" regarding the legal and regulatory framework, there are potential uncertainties regarding the clarification of possible conflicts between trading partners [10][35].

In particular, the 'Proof of Work' mechanism causes extremely high-energy consumption, which is a factor of several thousand higher than traditional financial transactions [36].

For a long-term success, a digital currency (using blockchain technology) must achieve the acceptance of the majority of the population. After all, the long-term importance of the digital currency ultimately depends on the number of actual users and the acceptance as a payment system by the trade [37].

E. Research Model – Adjusted Model with Elements of the Unified Theory of Acceptance and Use of Technology 2

In this section, the used research model will be described. The focus in this research paper will be on the relationships between (a) the risks of cryptocurrencies and the intention to use cryptocurrencies, (b) the experiences with cryptocurrencies and the intention to use cryptocurrencies, and (c) the general experiences with investments and the intention to use cryptocurrencies. The analysis of the named research concepts follows the Unified Theory of Acceptance and Use of Technology 2 (UTAUT2), which developed by Venkatesh, Thong, and Xin [38]. The UTAUT2 expands the existing UTAUT by the additional elements of hedonic motivation, price, and habit/experience, which allows a broader consideration of critical influence factors on the user behavior and the behavioral intention to use [38]-[40].

For this reason, to estimate these and further relations, an adjusted approach of the Unified Theory of Acceptance and Use of Technology 2 (UTAUT2) will be used, which is displayed in the conceptual model in Figure 1.

In principle, it can be assumed that higher returns or expected returns are generally associated with a higher investment risk. In this respect, it is necessary to examine how much risk they are prepared to take in order to achieve high returns. It can be assumed that investors who have more experience with investments and who have often made these via digital channels (e.g., online banking) are generally more open to the use of cryptocurrencies.

Finally, it should be noted that so far there has been no scientific review of the relationship between (a) performance expectations, (b) experience, (c) perceived risk and behavioral intention to use cryptocurrencies. The variable of perceived risk is treated as external variable in the further analysis. Additionally, the strength of perceived risk and experience will be estimated by linking these variables with the performance expectancy. The performance of the investments in digital currencies is rated by the performance expectancy.

Problematically, (a) the expected performance, (b) experience, and (c) perceptions of risks differ between the individual customers [41]. This means, the user attitudes and beliefs are completely subjective [41]. The experience comes from the fact that users become more and more familiar with a technology or service after it has been used for the first time. [39][42][43]. With the increasing use of a technology or a service, the user gains more and more experience and knowledge and learns with it, whereby the use becomes more and more self-evident and "automatic" [44].

Since habits and experiences allow predictions to be made for later use, it can be predicted that experience positively influences the utilization of cryptocurrencies.

In principle, the existing risks influenced the uses and investment behavior of customers [45]. This is particularly reflected in the fact that the risk has a significant influence on customer acceptance of innovations (e.g., mobile payment, mobile banking and mobile shopping) [46]-[51].

Previous researches identified that the perceived risk is one of the key drivers for the estimation of uncertainties in mobile payments, mobile shopping, mobile banking, and mobile transactions [46][48]-[52], because customers fear a lack of control.

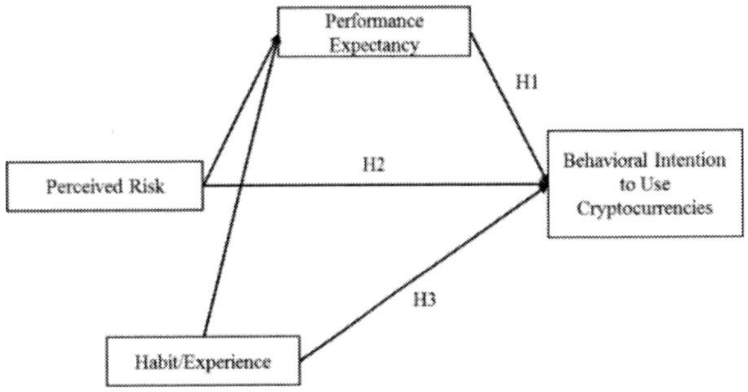

Figure 1. Conceptual Model

Consequently, the literature conveys the feedback that in several cases risks and uncertainties influence the customer user behavior.

Based on the explanations, the hypotheses for this research paper are:

H1: The customer perception of performance expectancy of investments (including digital currencies) has a directly positive effect on the intention to use cryptocurrencies.

H2: Customers' experience with investments (including digital currencies) has a direct positive impact on their intention to use cryptocurrencies.

H3: If customers are generally affine to risk when making investments, this has a direct positive effect on the intention to use cryptocurrencies.

III. Methodology

In this section, the approach for the verification of the hypotheses will be shortly introduced. Therefore, to test the hypotheses, an online survey was carried out to obtain information on the investment behavior of private individuals. In particular, the survey covered the perception of users regarding cryptocurrencies and the resulting investment behavior. The focus here is on the perceived performance of digital currencies and investments made.

To achieve the needed user information, a cross-sectional online survey ("one-shot survey") had been prepared and distributed through multipliers in social media platforms [53].

As this is an online survey, it cannot be guaranteed (as opposed to a personal survey) that most respondents will fully answer the questions. In addition, the questionnaire was designed in such a way that individual questions could not be skipped without ending the survey. In this respect, a relatively large number of participants prematurely terminated their responses to the questionnaire. The survey was distributed during the period from May to June 2018. In this period, 155 people have opened and started the questionnaire. However, only 62% (96 out of 155) respondents have finished the questionnaire. For this reason, the sample of the whole analysis will be the data set covering the 96 respondents, which have fully completed the questionnaire.

In the first part, the demographic information (such as age, gender and educational level) of the respondents was collected. In the following part the previous investment behavior and the knowledge of the participants about cryptocurrencies was determined. It should be determined whether the respondents know cryptocurrencies and whether they have already made investments based on cryptocurrencies. A positive answer (= experience with cryptocurrencies) was used to determine in more detail how many transactions, how much with which cryptocurrency the participants have already carried out. Since the third part is in higher importance for the later data analysis, all the implemented questions were coded in the 5-Point-Likert-scale format [54]. The third part covers especially questions regarding the respondent investment intentions in cryptocurrencies. In addition, the risk appetite and expected return (5-Point-Likert-scale: high to low) are important information in this part. The subsequent fourth part of the questionnaire considers questions regarding the user perception about the course of the cryptocurrency investment. As in the part before, the questions are coded in 5-Point-Likert-scale format (very likely to very unlikely). In the last part of the questionnaires, the respondents were queried about the future of cryptocurrencies in general.

The collected data were analyzed using quantitative research methods and the SPSS statistical program. To examine the reliability and validity of the data, the estimation of the Cronbach's Alpha and the Exploratory Factor Analysis were performed.

Only the eight largest cryptocurrencies (measured by market capitalization) were taken into account.

As mentioned above, the used approach only contains elements of the UTAUT2. Therefore, the evaluation is not done by structural equation modeling [38]. Instead, an ordinary least square regression to test the significance of each hypotheses is used. In the final hypothesis, all the previous considered single variables, like (a) perceived performance, (b) experience, (c) risk appetite, (d) investment and speculation type, (e) regulations, and (f) assessment of the acceptance as alternative payment method will be related to the undertaking of investments in cryptocurrencies.

IV. Data Analysis and Results

Following the described approach in conducting the survey, the outcomes for the estimation of the hypotheses will be deeply illustrated.

A. *Descriptive Results*

In the following, the descriptive results of the performed survey will be introduced. 53.1% of the respondents are male and the average age of a respondent is between 25 and 29 years. With 41.7% respectively 24.0%, the group of the 18 to 24 year respectively 25 to 29-year-old persons have the largest proportions of respondents within the survey. On the base that the age group of the 20 to 29 year old persons has only a 12.2% share of the total population in Germany, it must be noted that the young persons below the age of 30 years old are overrepresented in the survey by a factor of approx. five [55]. Since cryptocurrencies are virtual goods, their use requires a high Internet affinity. Based on a study of ARD/ZDF from 2015 the age group of the 20 to 29-year-old persons does nearly complete use the internet [56].

Since younger people generally use the Internet more often and have a greater interest in virtual goods than older people have, the previously established overrepresentation of younger age groups is not surprising. With regard to the age, the survey is not representative for the total population of Germany.

Considering the educational background, nearly the half of the respondents (46.9%) state that the school leaving examination is the education degree what they have. A quarter of the respondents have completed the Master degree (25%) from university.

The average net household income of the respondents is between 1,000 and 1,999 euros per month, with most of the participants (36.3%) having a net (household) income of less than 500 euros per month. In addition, almost three quarters (73.8%) have a net (household) income of less than 2,000 euros. In connection with the level of education and age, it can be assumed that the interviewees are predominantly students.

90.6% (= 87/96) of the respondents know what cryptocurrencies are. These 87 persons are the basic population (= 100%) for questions about cryptocurrencies.

47.9% (= 46/96) of the respondents have already made financial investments. However, only 35.4% (34 of 96 respondents) have already done investments in or with cryptocurrencies. From this point of view, the 34 respondents will be the basic population (= 100%) for all questions regarding the investment behavior with cryptocurrencies, especially number of transactions, amount of invested financial resources and perceptions regarding the development of the invested portfolio.

Firstly, the descriptive results for the respondents, who know cryptocurrencies (=87), will be illustrated. In general, all the respondents know Bitcoin as cryptocurrency, whereas two thirds of the respondents answer to know BitcoinCash and Ethereum, which can be seen in Table I.

TABLE I. DEGREE OF AWARENESS OF CRYPTOCURRENCIES

Cryptocurrency	Degree of Awareness
Bitcoin	**100.0%**
Bitcoin Cash	**67.1%**
Ethereum	**66.7%**
Litecoin	**61.4%**
Ripple	**58.5%**
EOS	**45.8%**
Neo	**41.0%**
Cardano	**35.4%**

Although 36.9% of the respondents are very risk-affine with regard to investments, only 23.4% of the respondents describe themselves as speculators. Contrary, 39.3% of the respondents answer to have a risk-shy nature, which can be also seen in estimation that 37.8% of the respondents estimate to be arbitragers. By regarding the estimation of returns, only 21.7% of the respondents think to get low returns. Although it is well known that higher returns can only be achieved with higher risks, some of the respondents who are risk-averse hope for medium to high returns.

Interestingly, 87.8% of respondents think that the new cryptocurrencies have been brought to life to drive a new form of speculation and investment. This is underlined by the fact that only 22.9% of respondents see Bitcoin as an alternative payment method to credit cards and the like. 43.4% of respondents involved in investment argue for regulatory intervention or restrictions in the cryptocurrency market, while 32.5% reject it.

Now, the results of the respondents using cryptocurrencies are shown. As already mentioned, however, the sample size is very small with 34 respondents, which is why the results cannot be generalized.

Figure 2 shows the distribution of the investment in the eight most important cryptocurrencies. 83.0% of respondents have already invested in Bitcoin. In addition to Bitcoin, the currency Ethereum seems to be of particular interest to investors.

48.5% of the respondents have invested at least 2,000 Euro in cryptocurrencies. 67.6%, these investors state that they make a profit by investing in cryptocurrencies. Figure 3 shows the objectives of the investments.

In general, most investors in cryptocurrencies believe in long-term increases in value. In comparison to the overall group of respondents knowing and using cryptocurrencies, the users of cryptocurrencies believe in a higher degree that Bitcoin could develop to an alternative currency and payment method.

In general, over 50% of the investors have a long-term direction by investing in cryptocurrencies.

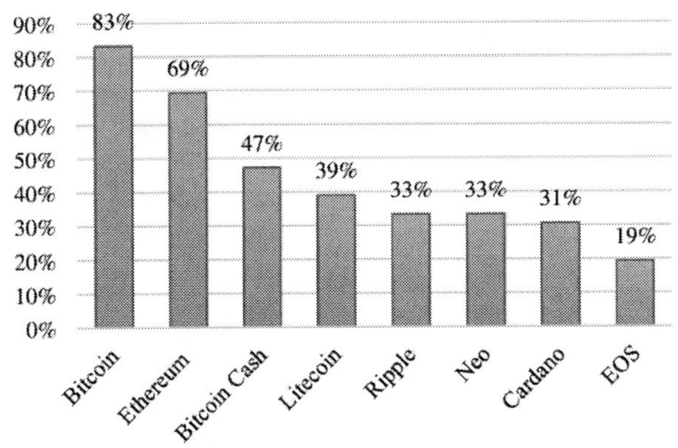

Figure 2. Investments in Cryptocurrencies

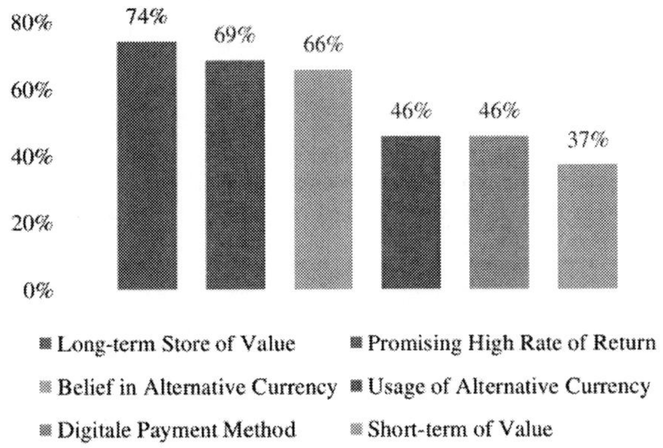

Figure 3. Purpose in Cryptocurrencies

In this respect, the investment in these currencies usually takes place with a longer time horizon (of several years).

Figure 4 illustrates the expectations of investors with regard to the performance of their currency investments.

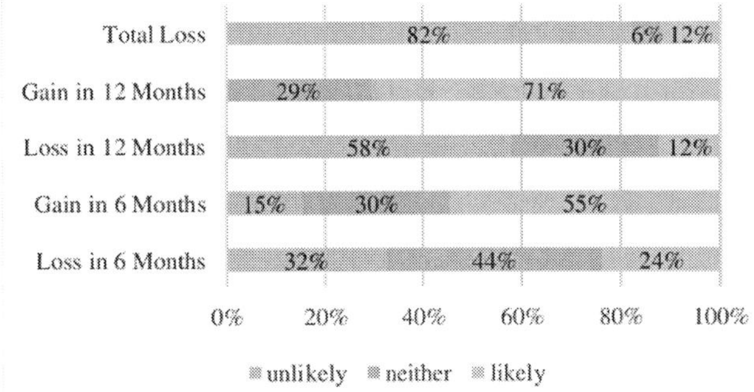

Figure 4. Expectations regard to the Cryptocurrency Investment

The majority of the investors (82%) assumes that a total loss of the investment does not occur. Contrary, over the half of the investors assumes to gain profits (in 6 months: 55%; in 12 months: 71%). Considering, the probability to get a loss in the investments, 58% of the investors estimate this as unlikely within the next 12 months.

Overall, the investment atmosphere regarding cryptocurrencies is quite positively. Investors perceive high profits by doing the investments and see only minor risks of a loss of their investments.

B. Reliability and Validity

The results of the reliability and validity analyses are illustrated in the Tables II and III. In general, this study includes the following 7 concepts: (1) performance expectancy, (2) experience, (3) perceived risk, (4) intention to use cryptocurrencies, (5) purposes of investments in cryptocurrencies, (6) usage of cryptocurrencies, and (7) prominence of cryptocurrencies.

TABLE II. RELIABILITY ANALYSIS

Research Concepts	Cronbach's Alpha
Performance Expectancy	**0.335**
Experience	**0.282**
Perceived Risk	**0.779**
Intention to Use Cryptocurrencies	**0.624**
Purposes of Investments in Cryptocurrencies	**0.777**
Usage of Cryptocurrencies	**0.726**
Prominence of Cryptocurrencies	**0.947**

Generally, all named concepts are examined in the terms of reliability and validity. Following Cronbach, Alpha values must be higher than 0.7/0.6 to for a good/sufficient reliability [57]-[59]. Based on the results in Table II, the collected data for 5 of the 7 named aspects are at least sufficiently reliable. Solely, the concepts of performance expectancy and experience seem to be completely not reliable.

After the testing of the reliability, the exploratory factor analysis includes the assessment of Kaiser-Meyer-Olkin criterion (KMO), the significance test from Bartlett, and the examination of the cumulative variance to evaluate the validity of the collected data [60]-[64]. Validity considers the consistency of an empirical measurement with the based conceptual/logical measurement concept. To reach a good validity, the concepts should reach significant p values ($p<0.05$) in the Bartlett-Test and KMO values above 0.7 (at least higher than 0.5) [60]-[64].

Table III shows a sufficient validity for 6 of the 7 concepts. The validity scores are also supported by the results of the cumulative variances higher than 50% except the concept of experience.

TABLE III. VALIDITY ANALYSIS

Research Concepts	KMO	Bartlett-Test	Cumulative Variance
Performance Expectancy	**0.284**	p < 0.000	**78.844%**
Experience	**0.562**	p < 0.000	**47.657%**
Perceived Risk	**0.637**	p < 0.000	**71.614%**
Intention to Use Cryptocurrencies	**0.640**	p < 0.000	**64.520%**
Purposes of Investments in Cryptocurrencies	**0.686**	p < 0.000	**74.544%**
Usage of Cryptocurrencies	**0.642**	p < 0.000	**69.377%**
Prominence of Cryptocurrencies	**0.911**	p < 0.000	**74.529%**

Despite the mark of 50% is not completely achieved, the explanatory rates of the variances can be rated as sufficiently high [61]-[63]. Consequently, the reliability and validity of the collected data are proved.

Table IV shows the variables that have a significant correlation with the intention to use cryptocurrencies. In addition to the values shown in Table IV: (a) There are positive significant correlations for all variables of perceived risk and experience with the intention of using cryptocurrencies.

C. Correlation Analysis

The correlation analysis measures the degree of the relationship between two individual variables. It is not, however, the degree of the linear proportionality. A correlation of 1.000 shows a 'perfect' relationship. A correlation coefficient higher than 0.500 is classified as a good correlation. Below 0.300, the correlation coefficients are weak [65][66].

TABLE IV. SIGNIFICANT CORRELATIONS FOR THE INTENTION TO USE CRYPTOCURRENCIES

Variables	Correlation Coefficient
Total Loss	**-0.349**
Risk Appetite	**0.475**
Expected Returns	**0.479**
Risk Type	**0.513**
General Investment	**0.728**
Investment Duration	**0.388**
Year of First Investment	**0.508**

(b) From the concept of performance expectancy, only the variable of the expectation regarding the total loss of an investment in cryptocurrencies correlates negatively significant with the intention to use cryptocurrencies.

D. Regression Analyses

As introduced earlier, the regression analysis will be performed on the method of an ordinary least square regression. The intension is to verify if the dependent variable behavioral intention to use cryptocurrencies is affected by the developed three concepts of independent variables [66]. In this regard, it will be examined, in which degree the predictor variables can explain the generated values of the dependent variable [67].

In the application of the regression analysis, four major indicators need to be considered. Firstly, the r-square will be determined to quantify the explanatory power of the whole regression model. The r-square is the share of the dependent variable, which can be explained by the independent variables. Following Chin and Cohen, the value should be at least 33% [68][69].

Secondly, the analysis of the variances (ANOVA) needs to verify the model fit. The resulting values should be significant ($p<0.05$) and higher than 3 in order to evaluate the model as good, which is the case here.

Thirdly, the regression coefficients of the independent variables need to be significant ($p<0.05$). In particular, the identified estimators must match the expectations in the research hypotheses. Fourthly, the test of multicollinearities by the Variance Inflation Factor (VIF) needs to be performed to find out, whether the variables included in the regression analyses have an identical relation. In the case of existing multicollinearities, i.e., if the VIF values exceed 10 (or in a stricter definition 3), the outcomes of the regression analysis are biased [60][70][71].

In performing the regression analysis, the relation between the variables of perceived risk and the intention to use cryptocurrencies are investigated (see Table V). In general, the r-square achieves a score of 35.2%. Since this value is slightly above the mark of 33%, there is at least a sufficient explanatory rate of the values of the dependent variable. The ANOVA scores an F-Ratio above the mark of 3.90.

The expected return positively significant affects the intention to use cryptocurrencies. This means, when investors expect a higher return, they are more open to use cryptocurrencies. In addition, the affinity for risk relates positively significantly but weakly with the intention to use cryptocurrencies. If investors are open to speculate and to take higher risks in investments, they intent to use cryptocurrencies for their investments. The VIF-values are below the mark of 3, so it can be excluded that multicollinearities are within the assumed model.

In Table VI, the variable of the expectation regarding a total loss of the investment is related to the intention to use cryptocurrencies. The r-square of the regression is 12.2%. Surely, the mark is below 33% and therefore, the explanatory rate seems to be low. In comparison to the other concepts, the expectation of a total loss of an investment in cryptocurrencies reaches a high r-square regarding that only one variable in the regression is considered. The F-Ratio of the ANOVA indicates a value better than the mark of 3.90 and therefore, a model fit is given. The variable total loss is negatively significant related to the intention to use cryptocurrencies.

TABLE V. REGRESSION ANALYSIS – PERCEIVED RISK:

Independent variables	**Dependent: Intention to Use Cryptocurrencies**	
ANOVA = 13.932	**R-Square = 35.2%**	
p<0.05	**Regression Coefficients with Significance**	**VIF**
Risk Appetite	**0.052**	**2.656**
Expected Return	**0.143****	**1.328**
Risk Type	**0.182***	**2.524**

* Significant within the error probability of 10%.
** Significant within the error probability of 5%.

TABLE VI. REGRESSION ANALYSIS – PERFORMANCE EXPECTANCY

Independent variables	**Dependent: Intention to Use Cryptocurrencies**	
ANOVA = 4.434	**R-Square = 12.2%**	
p<0.05	**Regression Coefficients with Significance**	**VIF**
Total Loss	**-0.057****	**1.000**

* Significant within the error probability of 10%.
** Significant within the error probability of 5%.

The negative relationship remarks that investors perceive that investments in cryptocurrencies are very improbable to lead to a full loss of the investment. This induces the openness for and investments in cryptocurrencies. Since there is only one variable, there cannot be any multicollinearities.

In Table VII, the variables of the concept experience are directly related to the intention to use cryptocurrencies. The r-square of 45.8% describes a low to moderate explanatory rate of the values occurring by the dependent variables. At least two fifths of the values of the dependent variable intention to use cryptocurrencies can be explained by applying the independent variables covering the concept of experience. The F-Ratio of 8.734 remarks an existing model fit.

TABLE VII. REGRESSION ANALYSIS – EXPERIENCE

Independent variables	**Dependent: Intention to Use Cryptocurrencies**	
ANOVA = 8.734	**R-Square = 45.8%**	
p<0.05	**Regression Coefficients with Significance**	**VIF**
General Investment	**0.365****	**1.090**
Investment Duration	**0.038****	**1.061**
Year of First Investment	**0.034**	**1.134**

* Significant within the error probability of 10%.
** Significant within the error probability of 5%.

In the concept experience, two variables are positively significant with the intention to use cryptocurrencies. Firstly, the general investment behavior positively affects to the intention to use cryptocurrencies. In general, in case investors do regularly investments (indifferently in which field) they are more open to intent to use investments in cryptocurrencies. Secondly, the variable, which includes the investment duration, is positively significant related to the intention to use cryptocurrencies. This means, investors are more oriented in a long-term store of value. If they behave in this direction, they see cryptocurrencies also as opportunity to invest over a longer time. If investors want to invest for a longer period of time, they more intent to use cryptocurrencies for their investments. The VIF-scores identify that multicollinearities can be excluded in the model.

Finally, in a combined regression, all independent variables of the three individual regressions are used together. The combination of the independent variables leads to an enhancement to the level of 70.1%. Comparing the resulting r-square to the mark of 33%, the combined approach identifies a high level of explanatory power. In this regard, nearly three quarters of the data points of the dependent variable can be explained by the application of the independent variables. The F-Ratio of 8.041 identifies a good model fit. Through combining all independent variables of the previous regression analyses, only the variable covering the general investment behavior affects positively significant the intention to use cryptocurrencies. When investors have more experience with the application and execution of investments in general, they are more open and willing to use cryptocurrencies. This effect seems to be the most dominant one in the model, since all the other independent variables are getting insignificantly when they are considered in the combined approach. It can be assumed that investors in cryptocurrencies are persons, who have done investments in the past. Therefore, if persons are familiar with investments, they are more willing to do investments in cryptocurrencies.

However, the combined approach identifies two variables (risk appetite and risk type), which have VIF-values above the mark of 3. In this regard, the combined approach cannot fully guarantee that no multicollinearities are within the model. For this reason, the regression coefficients could be biased by the overwhelming effects of the independent variables, which are strongly correlating with each other.

V. Conclusions And Future Work

Summarizing the regression analyses, the hypotheses H1 to H3 can be accepted. In general, when investors have made investments in the past, they are more open to use cryptocurrencies. This result is supported by the fact that how longer the investors do investments and have a long-term store of value, they intent to use cryptocurrencies. In addition, if the investors expect to experience not a total loss of the investment in cryptocurrencies, they have a higher willingness to use cryptocurrencies. Lastly, investors, who do investments with a greater risk, they have also a greater intent to use cryptocurrencies for their investment to reach higher returns.

To sum up, all three concepts identify significant variables, which are influencing the intention to use cryptocurrencies. For this reason, the assumed research model and hypotheses can be fully confirmed. However, as remarked in the beginning, the sample size of the whole analysis is too low. On this account, the achieved results cannot be generalized, and further quantitative analyses and surveys are necessary to deepen the influence factors of cryptocurrencies. As this is a very topical issue, the authors expect that further research works will be performed, which focus on the influence factor for the adoption of Bitcoin, Ethereum and further currencies.

References

[1] CoinMarketCap, "Top 100 Cryptocurrencies," 2019. [Online]. Available from: https://coinmarketcap.com/de/ [retrieved: 09.2019]

[2] F. Glaser, K. Zimmermann, M. Haferkorn, M. C. Weber, and M. Siering, "Bitcoin - Asset or Currency? Revealing Users," Hidden Intentions, Twenty Second European Conference on Information Systems, Tel Aviv, pp. 1-14, 2014.

[3] E. Sixt, "Bitcoins and further Transaction Systems. Blockchain as Base for Crypto Economy," [German] "Bitcoins und andere dezentrale Transaktionssysteme. Blockchains als Basis einer Kryptoökonomie," Wiesbaden, Springer Gabler, 2017. [Online]. Available from: http://dx.doi.org/10.1007/978-3-658-02844-2 [retrieved: 09.2019]

[4] P. Vigna and M. Casey, "The Age of Cryptocurrency. How Bitcoin and Digital Money are challenging the Global Economic Order," New York, NY, St. Martin's Press, 2015.

[5] S. Nakamoto, "Bitcoin: A Peer-to-Peer Electronic Cash System," 2008 [Online]. Available from: https://bitcoin.org/bitcoin.pdf [retrieved: 09.2019]

[6] J. Schütte et al., "Blockchain and Smart Contracts – Technology, Research and Applications," [German] "Blockchain und Smart Contracts – Technologien, Forschungsfragen und Anwendungen," 2017. [Online]. Available from: https://www.fraunhofer.de/content/dam/zv/de/forschung/artikel/2017/FraunhoferPositionspapier_Blockchain-und-Smart-Contracts_v151.pdf [retrieved: 09.2019]

[7] T. Aste, P. Tasca, and T. Di Matteo, "Blockchain Technologies: The Foreseeable Impact on Society and Industry," IEEE. 2017, ISBN: 0018-9162/17, pp. 18-28, 2017.

[8] O. Nica, K. Piotrowska, and K. R. Schenk-Hoppé, "Cryptocurrencies: Economic benefits and risks," FinTech working paper, University of Manchester, Manchester, pp. 1-56, 2017. [Online]. Available from: https://papers.ssrn.com/sol3/papers.cfm?abstract_id=3059856 [retrieved: 10.2019]

[9] S. R. Basnet and S. Shakya, "BSS: Blockchain Security over Software Defined Network," International Conference on Computing, Communication and Automation (ICCCA2017), IEEE, pp. 720-725, 2017.

[10] D. López and B Farooq, "A Blockchain Framework for Smart Mobility," IEEE Xplore Digital Library, 2018 IEEE International Smart Cities Conference (ISC2), ISBN: 978-1-5386-5959-5/18, 2018.

[11] S. G. Sharma and L. Ahuja, "Building Secure Infrastructure for Cloud Computing using Blockchain," IEEE Xplore Digital Library, 2018 Second International Conference on Intelligent Computing and Control Systems (ICICCS), ISBN: 978-1-5386-2842-3/18, 2018.

[12] R. Henry, A. Herzberg, and A. Kate, "Blockchain Access Privacy: Challenges and Directions," IEEE Computer and Reliability Societies, *IEEE Security & Privacy*, vol. 16, issue 4, ISBN: 1540-7993/18, pp. 38-45, 2018.

[13] C. Richter, "Blockchain Fog lifts for Assurance," [German] "Der Blockchain-Nebel lichtet sich auch für die Assekuranz - Vom Hype zum Geschäftsmodell für Versicherer," 2017. [Online]. Available from: https://www.accenture.com/de-de/_acnmedia/PDF-55/Accenture-Der-Blockchain-Nebel-Assekuranz-German-2017.pdf [retrieved: 09.2019]

[14] P. Urien, "Blockchain IoT (BIoT): A New Direction for Solving Internet of Things Security and Trust Issues," IEEE, 2018 3rd Cloudification of the Internet of Things (CIoT), 2018.

[15] S. Wang, L. Ouyang, Y. Yuan, X. Ni, X. Han, and F.-Y. Wang, "Blockchain-Enabled Smart Contracts: Architecture, Applications, and Future Trends," IEEE Transactions on Systems, Man, and Cybernetics Systems, ISBN: 2168-2216/2019, pp. 1-12, 2019.

[16] A. M. Antonopoulos, "Mastering Bitcoin: Unlocking Digital Cryptocurrencies," O'Reilly Media, 2014.

[17] G. W. Peters and E. Panayi, "Understanding Modern Banking Ledgers through Blockchain Technologies. Future of Transaction Processing and Smart Contracts on the Internet of Money," University College London, London School of Economics. London, pp. 1-33, 2015.

[18] V. Gatteschi, F. Lamberti, C. Demartini, C. Pranteda, and V. Santamaria, "To Blockchain or Not to Blockchain: That is the Question," IEEE Computer Society, *IT Professional*, vol. 20, issue 2, ISBN: 1520-9202/18, pp. 62-74, 2018.

[19] D. Fakhri and K. Mutijarsa, "Secure IoT Communication using Blockchain Technology," IEEE Xplore Digital Library, 2018 International Symposium on Electronics and Smart Devices (ISESD), ISBN: 978-1-5386-6670-8/18, 2018.

[20] C. O'Flynn, "A Lightbulb Worm? Details of the Philips Hue SmartLighting Design," Black Hat USA 2016.

[21] SRC – SIA/SRC, "Rebooting the IT Revolution: A Call to Action," 2015.

[22] E. Ronen, C. O'Flynn, A. Shamir, and A. Weingarten, "IoT Goes Nuclear:Creating a ZigBee Chain Reaction," IEEE Symposium on Securityand Privacy, *IEEE Security & Privacy*, vol. 16, issue 1, pp. 54-62, 2017.

[23] A. Back, "Hashcash. A Denial of Service Counter-Measure," 2002. [Online]. Available from: http://www.cypherspace.org/hashcash/ [retrieved: 09.2019]

[24] Microsoft, "Understanding Public Key Cryptography," 2014. [Online]. Available from: https://docs.microsoft.com/en-us/previous-versions/tn-archive/aa998077(v=exchg.65) [retrieved: 09.2019]

[25] M. I. Harrigan, L. Shi, and J. Illum, "Airdrops and Privacy: A Case Study in Cross-Blockchain Analysis," IEEE Xplore Digital Library. 2018 IEEE International Conference on Data Mining Workshops (ICDMW), pp. 63-70, 2018.

[26] K. Christidis and M. Devetsikiotis, "Blockchains and Smart Contracts forthe Internet of Things," IEEE Access, vol. 4, pp. 2292–2303, 2016.

[27] R. Palkovits, N. Pohlmann, and I. Schwedt, "Blockchain Technology revolutionize Digital Business," [German] "Blockchain-Technologie revolutioniert das digitale Business. IT-Sicherheit," pp. 54-60, 2017.

[28] M. Schneekluth, "Blockchain Consensus" [German] "Blockchain Konsens: Welche Konsens-Algorithmen gibt es?," 2018. [Online]. Available from: https://www.wallstreet-online.de/nachricht/10452672-blockchain-konsens-algorithmen-blockchain-konsens-konsens-algorithmen-es/all [retrieved: 09.2019]

[29] J. Altmann, "250 Keywords Bank Economy," [German] "250 Keywords Bankwirtschaft: Grundwissen für Fach- und Führungskräfte," Wiesbaden, Germany, Springer-Verlag, 2018.

[30] S. Haber and W. S. Stornetta, "How to Time-Stamp a Digital Document," in: Journal of Cryptology, vol. 3, pp. 99–111, 1991. [Online]. Available from: https://link.springer.com/article/10.1007%2FBF00196791 [retreived 09.2019]

[31] T. Wu and X. Liang, "Exploration and Practice of Inter-bank Application Based on Blockchain", The 12th International Conference on Computer Science & Education, University of Houston, USA, pp. 219-224, 2017.

[32] M. Rosenberg, "Bitcoin to Smart Contract – Use Cases," [German] "Von Bitcoin zum Smart Contract - Anwendungspotenziale der Blockchain-Technologie," 2016. [Online]. Available from: https://www.bmwi.de/Redaktion/DE/Downloads/Monatsbericht/Monatsbericht-Themen/10-2016-bitcoin.pdf?__blob=publicationFile&v=7 [retrieved: 09.2019]

[33] H. U. Buhl, A. Schweizer, and N. Urbach, "Blockchain Technology as Key for the Future," [German] "Blockchain-Technologie als Schlüssel für die Zukunft?," in: Zeitschrift für das gesamte Kreditwesen : Pflichtblatt der Frankfurter Wertpapierbörse, vol. 70, issue 12, pp. 596–599, 2017.

[34] A. Hayes, "Decentralized Banking: Monetary Technocracy in the Digital Age," in Banking Beyond Banks and Money, pp. 121–131, 2016.

[35] USSEC – U.S. Securities and Exchange Commission, "Investor Bulletin: InitialCoin Offerings," 2018. [Online]. Available from: https://www.sec.gov/oiea/investor-alerts-and-bulletins/ib_coinofferings [retrieved: 09.2019]

[36] M. Conti, S. Kumar, C. Lal, and S. Ruj, "A Survery on Security and Privacy Issues of Bitcoin, " *IEEE Communications Surveys & Tutorial*, vol. 20, issue 4, pp. 3416-3452, 2018.

[37] A. J. Schwartz, "Money Supply," in Concise Encyclopedia of Economics, 2008. [Online] Available from: http://www.econlib.org/library/Enc/MoneySupply.html [retrieved: 09.2019]

[38] V. Venkatesh, J. Y. L. Thong, and X. Xin, "Consumer Acceptance and Use of Information Technology: Extending the Unified Theory of Acceptance and Use of Technology," *MIS Quarterly*, vol. 36, issue 1, pp. 157-178, 2012.

[39] V. Venkatesh, M. G. Morris, G. B. Davis, and F. D. Davis, "User Acceptance of Information Technology: Toward a Unified View," *MIS Quarterly*, vol. 27, issue 3, pp. 425-478, 2003.

[40] F.-T. Lin, H.-Y. Wu, and T. T. Nguyet Nga, "Adoption of Internet Banking: An Empirical Study in Vietnam," 10th International Conference on e-Business Engnineering, IEEE Xplore Digital Library, pp. 282-287, 2013.

[41] S. Dhawan, K. Singh, and S. Goel, "Impact of Privacy Attitude. Concern and Awareness on Use of Online Social Networking," 5th International Conference - Confluence The Next Generation Information Technology Summit 2013, IEEE Xplore Digital Library, pp. 14-17, 2014.

[42] S. S. Kim and N. K. Malhotra, "A Longitudinal Model of Continued IS Use: An Integrative View of Four Mechanisms Underlying Post-Adoption Phenomena," *Management Science*, vol. 51, issue 5, pp. 741-755, 2005.

[43] S. S. Kim, N. K. Malhotra, and S. Narasimhan, "Two Competing Perspectives on Automatic Use: A Theoretical and Empirical Comparison," *Information Systems Research*, vol. 16 (4), pp. 418-432, 2005.

[44] M. Limayem, S. G. Hirt, and C. M. K. Cheung, "How Habit Limits the Predictive Power of Intentions: The Case of IS Continuance," *MIS Quarterly*, vol. 31, issue 4, pp. 705-737, 2007.

[45] T. Zhou, "Understanding Mobile Internet Continuance Usage from the Perspectives of UTAUT and Flow, " *Information Development*, vol. 27, pp. 207-218, 2011.

[46] J. Zhong, A. Dhir, M. Nieminen, M. Hämäläinen, and J. Laine, "Exploring Consumer Adoption of Mobile Payments in China," Academic Mind Trek'13, pp. 318-325, 2013.

[47] Y. S. Wang, Y. M. Wang, H. H. Lin, and T. I. Tang, "Determinants of User Acceptance of Internet Banking: an Empirical Study," *International Journal of Service Industry Management*, vol. 14, pp. 501-519, 2003.

[48] L.-D. Chen, "A Model of Consumer Acceptance of Mobile Payment," *International Journal of Mobile Communications*, vol. 6, issue 1, pp. 32-52, 2008.

[49] M. A. Mahfuz, L. Khanam, and W. Hu, "The Influence of Culture on M-Banking Technology Adoption: An Integrative Approach of UTAUT2 and ITM," 2016 Proceedings of PICMET'16: Technology Management for Social Innovation, pp. 70-88, 2016.

[50] X. Luo, H. Li, J. Zhang, and J. P. Shim, "Examining Multi-dimensional Trust and Multi-faceted Risk in Initial Acceptance of Emerging Technologies: an Empirical Study of Mobile

Banking Services," *Decision Support Systems*, vol. 49, issue 2, pp. 222-234, 2010.

[51] H.-P. Lu and P. Y.-J. Su, "Factors Affecting Purchase Intention on Mobile Shopping Websites," *Internet Research*, vol. 19, issue 4, pp. 442-458, 2009.

[52] T. Zhou, "An Empirical Examination of Initial Trust in Mobile Banking," *Information Development*, vol. 21, issue 5, pp. 527-540, 2011.

[53] A. Diekmann, "Empirical Social Research," [German] "Empirische Sozialforschung," Rowohlt-Taschenbuch-Verlag. Reinbek bei Hamburg, vol. 5, 2011.

[54] R. Likert, "A Technique for the Measurement of Attitudes," Archives of Psychology, pp. 199-224, 1932.

[55] Destatis, Statistisches Bundesamt, "Population," [German] "Bevölkerung," 2015. [Online]. Available from: https://www.destatis.de/DE/ZahlenFakten/GesellschaftStaat/Bevoelkerung/Bevoelkerungsstand/Tabellen_/lrbev01.html [retrieved: 09.2019]

[56] Statista, "Internet Users in Germany from 2001 to 2015," [German] "Anteil der Internetnutzer in Deutschland in den Jahren 2001 bis 2015," 2015. [Online]. Available from: http://de.statista.com/statistik/daten/studie/13070/umfrage/entwicklung-der-internetnutzung-in-deutschland-seit-2001/ [retrieved: 09.2019]

[57] L. J. Cronbach, "Coefficient Alpha and the Internal Structure of Tests," Psychometrika, vol. 16, pp. 297-334, 1951.

[58] C. Fornell and D. Larcker, "Evaluating Structural Equation Models with Unobservable Variables and Measurement Error," *Journal of Marketing Research*, vol. 18, issue 1, pp. 39-50, 1981.

[59] R. Hossiep, "Cronbachs Alpha," [German] "Cronbachs Alpha," In Wirtz, M. A. (editor): Dorsch – Lexikon der Psychologie, vol. 17. Verlag Hans Huber, Bern, 2014.

[60] J. F. J. Hair, R. E. Anderson, R. L. Tatham, and W. C. Black, "Multivariate Data Analysis," Macmillan, New York, NY, Macmillan, vol. 3, 1995.

[61] S. Fromm, "Data Analysis with SPSS Part 1," [German] "Datenanalyse mit SPSS für Fortgeschrittene," Arbeitsbuch, vol. 2, VS Verlag für Sozialwissenschaften, GWV Fachverlage, Wiesbaden, 2008.

[62] S. Fromm, "Data Analysis with SPSS Part 2," [German] "Datenanalyse mit SPSS für Fortgeschrittene 2: Multivariate Verfahren für Querschnittsdaten," Lehrbuch, vol. 1, VS Verlag für Sozialwissenschaften, Springer, Wiesbaden, 2010.

[63] N. M. Schöneck and W. Voß, "Research Project," [German] "Das Forschungsprojekt – Planung, Durchführung und Auswertung einer quantitativen Studie," vol. 2. Springer Wiesbaden, 2013.

[64] A. Field, "Discovering Statistics Using SPSS," Sage Publications Ltd., vol. 4, 2013.

[65] S. Hagl, "Statistics," [German] "Schnelleinstieg Statistik," Rudolf Haufe Verlag. München, vol. 1., 2008.

[66] F. Brosius, "SPSS 8 Professionell Statistics in Windows," [German] "SPSS 8 Professionelle Statistik unter Windows," Kapitel 21 Korrelation. International Thomson Publishing. edition. 1, 1998.

[67] T. Schäfer, "Stastics I – Descriptve Results and Explorative Data Analysis," [German] "Statistik I. Deskriptive und Explorative Datenanalyse," VS Verlag für Sozialwissenschaften, 2010.

[68] W. W. Chin, "The Partial Least Squares Approach for Structural Equation Modeling," in G. A. Marcoulides (Ed.). Modern Methods for Business Research. Lawrence Erlbaum Associates. Mahwah, NJ, pp. 295-336, 1998.

[69] J. Cohen, "Statistical Power Analysis for the Behavioral Sciences," Lawrence Erlbaum Associates. Hillsdale, edition 2, 1988.

[70] S. Petter, D. W. Straub, and A. Rai, "Specifying Formative Constructs in Information Systems Research," *MIS Quarterly*, vol. 31, issue 4, pp. 623-656, 2007.

[71] D. Lin, D. P. Foster, L. H. Ungar, "VIF Regression: a Fast Regression Algorithm for Large Data," *Journal of the American Statistical Association*, vol. 106, issue 493, pp. 232-247, 2009.

Personalizing Learning with m-Technologies to Improve Students Performance in South African Higher Learning

Refilwe Constance Mogase
Department of Informatics
Tshwane University of Technology
Pretoria, South Africa
mogaserc@tut.ac.za

Patricia M Alexander
Department of Informatics
Tshwane University of Technology
Pretoria, South Africa
alexandermp@tut.ac.za

***Abstract*— Personalized Learning (PL) is recently being endorsed by higher institutions of learning as a strategy for improving students' engagement and academic attainment. Although existing PL technologies offer some flexibility, there is still room for further innovation to extend this learning support. However, PL still lacks fully adaptive and timely content, which can enhance their flexibility further. Often times instructors are faced with a challenge of giving students attention in the undergraduate courses. This limitation is a consequence of fewer contacts sessions between lecturers and students due to timetable schedules. Therefore, it is imperative for institutions of higher learning to consider integrating PL through use of mobile devices to extend learning beyond the institution's geographical boundaries. Some theoretical foundations already exist for learning through personalisation and learner perceptions on using technologies to develop effective PL environments. However, there exists the need to develop models for mobile computing personalized learning in the African context to best suit the needs of students from the socio-economic backgrounds in support of their learning environments. The study aims to develop a mobile computing model to enhance PL for students in South African context with the aim of improving their academic performances. Content analysis was used to evaluate the factors that had been identified in the literature and were tallied according to their frequency. The results indicated that device factors that relates to technology use along with pedagogy predominantly impact on the effectiveness of PL using mobile technologies.**

Keywords- Mobile computing, Personalized learning, Technology-supported personalized learning environments, M-Learning

I. INTRODUCTION

Education in the 21s century takes into cognizance of the fact that students think differently, where their thinking is influenced by their social, cultural and technological environment [3]. This suggests that their learning is a constructive process of acting within an environment full of ubiquitous computing devices. These devices are supported by wireless communication networks, which make possible to access content anywhere anytime. Information and Communication Technology (ICT), which has presented opportunities for knowledge creation and easy access to information, has modified the conception of time and space in the learning environment. Mobile computing play an important role not only to education, but to society, and it is further making a positive contribution to other areas outside the learning spheres such as health, management, scientific research and business administration [23]. Mobile technologies lay a good foundation for expanding teaching and learning that goes beyond the confines of classroom and university campus. It further provides a platform that support student collaboration, sharing of information, and flexibility in learning if employed effectively, which can support social constructivist approaches to learning [7]. PL through use of digital resources has promoted new version of learning by tailoring learning to suit learner needs, where emphasis is on supporting the learner [21]. The combination of these two learning strategies can offer an impending educational innovation that is significant for the flexible delivery of learning content. This mix of learning strategies allows for the customization and personalization of student's learning process.

In essence, integrating mobile technologies and personalization in the formal learning will enable a learning environment that is ubiquitous where students may be leaning without even being fully aware. This will be due to the social and personal aspects of mobile devices, where the limits and distinctions between work and play, learning and entertainment, formal and informal learning are no longer relevant. This research will identify key components that contribute to the implementation of learning tailored for a specific set of learners. This takes into consideration student's learning needs, interests and preferences. More so, the study is seeking to make a contribution to higher education (HE) by combining both PL and mobile technologies to ensure that learning is made flexible, fun, and seamless with the aim of improving student's academic performances.

This study investigated the factors that impact on the effective implementations of PL using mobile technologies to improve student's learning and their academic performances. The paper was structured in a form of sections indicating a brief description of what each section entails. In Section 2, the study highlights PL background in the context HE and its impact on student's learning processes. Section 3 specifies the gaps that exists in PL environments along with the goal and objectives the study. Section 4 gives a brief literature review indicating PL in higher education, PL challenges and the opportunities it brings to academia, this is followed by related works indicating various factors that need to be taken into consideration to ensure the effectiveness of PL using mobile technologies in HE. A buildup of a conceptual framework that underpins the study is then described concluding the section. In Section 5, the methodology adapted for the study is explained in detail followed by a table of results that identifies the key factors that are sort key to the implementations of PL environments using mobile technologies. Section 6 provides future works and a conclusion of the study.

II. BACKGROUND

PL facilitates both formal and informal education since it supports collaboration and communication between the learner and his or her instructor and between learners. In addition, PL shifts the role of students from being simply consumers of education to co-producers and collaborators of their learning pathway [21]. Through use of technologies, the style of teaching and learning can be modified to suit a group of students with diverse backgrounds and experiences [1]. More so, institutions of higher learning are able to consider a variety of student's needs and characteristics within the teaching and learning process through personalization. The emphasis of PL is on the usage of flexible instructional practices in order to facilitate students' personalization endeavors.

PL can blend different forms of learning and techniques to cater for learners who learn in different ways. The focus of PL is on the choice of what to study, how and when, whilst considering the pace at which the student learn, tailored to students' learning preferences and specific interests of different learners. Bingham et al. [6] points out that PL, powered by technology, can allow for more productive time-on-task. However, its extensions are sometimes limited by lack of resources such as internet accessibility, PCs or laptops due to their cost.

Phillipson et al. [24] noted that currently PL lacks the mobility aspect that allows learning to reach higher pedagogical level, where constructing and absorbing knowledge becomes life long, breaking the boundaries of time and space. This can be achieved if PL is designed based on mobile computing technologies and through the use of portable devices such as Mobile phones, PDA, iPads and other devices built on mobile technology.

III. PROBLEM STATEMENT

There is an increased pressure on higher institutions of learning by higher education department to meet quantitative performance targets [11] which had led to a massive numbers of students enrolled in African Universities. Often times instructors are faced with a challenge of giving students attention in the undergraduate courses [26]. This limitation is a consequence of fewer contacts sessions between lecturers and students due to timetable schedules. More so, there is lack of instructional goals that involves high levels of cognitive skills, low self-esteem among learners, and minimum opportunities of individual feedback. While ICT has the potential to provide higher institutions with such learning opportunities, its use is still limited in the SA context. As a result, there is a demand for instructors to reach every individual student by incorporating ICT interventions such as mobile technologies. To recap, mobile tools have the inherent ability to connect with information tools and communicate in a networked manner and this makes them valuable in a situated learning context [11]. There is a need for constant connectivity in order for just-in-time learning and students' collective knowledge sharing to be leveraged to promote collaboration [3].

On the other hand, studies have built theoretical foundations that address learning through personalization and learner perceptions on using technologies to develop effective PLEs [19]. However, due to PL in formal learning being a relatively new technological enhancement, there seems to be few studies that have explored this area of research [4]. As a result, little guidance has been given to enhance interactions that support extended flexibility, mobility, reachability and accessibility for teaching and learning in PL environments [2]. Therefore, there is a need for personalized learning models that caters for a specific set of learners and their supporting environments.

To solve the highlighted problem, the following goal and research objectives have to be met.

1) Goal

To design a Mobile Computing model to enhance Personalized Learning for students in South African higher learning

2) Specific objectives

1. To determine and analyse the influence of PL to effective teaching and learning of students in higher learning institutions
2. To investigate and analyse technology-based PL models for higher learning institutions
3. To investigate the determinants that influence the incorporation of mobile computing for PL
4. To develop a mobile computing model that can be used to enhance PL for students in higher institutions

IV. LITERATURE REVIEW

This section provides an overview of PL in HE, its benefits and challenges. Furthermore, it previews works that have been done in light of presenting key factors that impact the effective implementations of PL in academic institutions of higher learning.

A. Personalized Learning in Higher Education

PL extends learning through a mix of pedagogies such as student mentoring, incorporation of learners' personal and social experience, using ICT, and other learning strategies such as flipped learning. This provides learner support which reforms learning from a 'fixed content and fixed timing' concept. Accounts of pedagogies relating to PL and ML overlap to some extent but reflect different ways of offering flexibility in the learning processes. However, their distinct difference is in the learning mode of delivery and information filtering to cater for individual needs [29]. While they both incorporate ICT tools to offer blended learning, ML with PL adds just-in-time, on-the-go learning, and easy access to PL resources due to its mobility and portability capabilities.

Ross et al. [25] supports this active learning strategy that puts the learner at the center of teaching and learning with students' independently exploring, generating, and applying ideas in and outside the classroom setting. These instructional methods have proven to show amicable improvements in student's conceptual understanding when correctly applied.

Waldrip and Vaughan [28] also highlighted note-worthy facts in their meta-analysis review of the effects of integrating mobile devices with teaching and learning on students' learning performance. They alluded to various scholars' resulting consensus that to exert maximum effect of learning with technologies in higher education institutions, there should be reconciliation of the connections between the following components: 1) technology (both hardware and software), 2) process of teaching and learning to accommodate different settings (educational context), 3) and users (both lecturers and students) [25].

B. Benefits of PL in Higher Institutions of Learning

The aim of PL is to provide a positive learner experience that is based upon information about the learner. Students have the opportunity to access services that are most relevant to their learning context which suit their interests saving them time to filter out from the vast amount of information available. PL's main objective is to address the intended 'one-size-fits-all' approach, provide support that is dependent on the student's engagement with the course material, and peer interactions [26].

Verdu et al. [27] alludes to some of the benefits that PL brings. These include increased motivation, learner empowerment, and improved attitudes to learning. PL further contributes towards sharing and developing expertise, through collaboration and networking. It produces better results through a student's personal development of better learning strategies, skills, and technological capacities. It caters for differences in learning contexts and cultures in terms of sociocultural practices in urban versus developing areas. More so, PL enable students to have the agency when making choices of what, when, how and where to learn, where control over setting their learning goals is within their reach. Learning is made flexible enough to allow students to take their learning outside the confines of the traditional classroom.

C. Challenges of Personalizing Learning

While the idea of personalization might be appealing, the reality of its implementation is much more complex. As Ross et al. [25] indicated many universities are faced with the dilemma of same content presentation in different sequence for different students thus not serving the purpose of filtering it to cater for a diverse set of students. Hostler [14] postulates the difficulties of implementing PL especially for larger groups. This is because not only does it require training but also time to accommodate the approach in the lecturer's teaching practices. In addition, it requires restructuring of curriculum and change of teacher classroom practices. The introduction of new technologies calls for support and on-going training to address specific challenges that may arise while using PL tools and spaces. More so, with large lecture environments it is difficult to individualize the learning to ensure all students' misconceptions are addressed. Therefore, careful thought and consideration should be put forth in terms of structuring learning when applying this PL strategy to have an effective lecture session.

D. Related Work

Laohajaratsang [17] explored the use of technology to personalize learning in a study for personalized, flexible blended learning features of the Moodle Learning Management System (LMS) through mobile platforms. The study factored the following as key to the implementation of PL, change in teaching practices including instructors skills, need to develop the learning content, support staff needed to manage, inspire and support both student and lecturers, and training particularly in areas of mobile content creation and pedagogical skills.

On the other hand Callaghan et al. [8] looked at using web-based technologies as pedagogy to enhance learning. In in their findings they indicated that students' academic performance increased by 67%. They emphasized the universities' need to explore other innovative ways beyond web-based technologies that could be used to ensure seamless learning environments. Furthermore, the study identified 8 factors that can enhance learning through use of technology. These included high-performance network infrastructure, choice of digital devices and software to support the pedagogical model underlying the course, content presentation to facilitate ease of use, social interaction and human touch, flexibility and adaptability of systems to cater for learner diversity, stimulating interaction, facilitating student's learning processes, and fostering an effective learning climate, specifically in the context of African higher institutions of learning.

Pimmer et al. [23] highlighted the works of Casany et al. [9], which presented a Protus system that was built on the foundations of an intelligent tutoring system. The system recognized different patterns of learning styles and learner habits through testing their learning styles and mining their server logs as key to ensure system flexibility. By so doing the requests done by students would then be customized to suit their learning preferences thus creating a learning space that is flexible and personalized to their needs.

More so, Pimmer et al. [23] emphasized on use of gaming for computing education as another means of personalizing learning for students. Its ability to teach specific concepts and strengthen particular competencies helps students to stay focused on the learning activity while stimulating their will to achieve. However, majority of evaluations on gaming education were ran with small samples and they lacked scientific rigor. Therefore, there is a need for further research on how to improve scientific rigor of their evaluations for teaching computing.

Studies by Verdú et al. [27] were carried out in the spheres of PL which highlighted the need for incorporating mobile technologies to extend learning flexibility. The study identified among other factors, the need for stimulating interaction, facilitating student's learning processes, and fostering affective learning climate.

E. Theoretical Foundations and Conceptual Model

Several theories have been adopted to underpin learning with technologies, particularly using mobile devices to support or extend teaching and learning, which came with their shortfalls [22] ;[6]. These theories lacked the inclusion of curriculum that is more tailored to student needs as an important aspect that would strengthen their personalized experiences. Instead, their focus was more on mobile technologies embedded in their current LMS. Studies of Bachman and Gannod [5] have used FRAME by Koole [16] aiming to explain mobile learning dynamics. They added additional characteristics that affected the device, learner, and social aspects in M-learning and referred to their conceptual model as augmented FRAME. Lai et al. [18] also used FRAME as a reference model to design tablet-based activities for pre-service teachers, which addressed aspects of collaboration, engagement, motivation and independent learning. Other studies by Juvova et al. [15] for example used and explained FRAME as an evaluation framework and showed through demonstrations how it could be applied in M-learning studies.

However, there is still some limitations in these theories in terms of the learner side when creating learning experiences that support individualized learning [23]. Moreover, various proportions need to be taken into consideration to help in making recommendations that could help in facilitating PL materials that can be administered via mobile devices. In addition, there is still lack of evaluation methods that are multi-attributed to help in modeling student needs for the creation of PL environments that more learner-centered [3].

The proposed study aims to reference two existing theoretical frameworks, FRAME by Koole [16] and a framework of Mobile Learning Preference by Lai et al. [18] to combine its variables to develop a pedagogical model for PL via MC for institutions of higher learning in South Africa. It is believed that no current framework can address the needs and interests of users in a holistic manner. The intent is to get insights based on the identified elements for PL through qualitative methods, which will help in determining whether they should be considered as key elements for effective design and implementation of a mobile computing model for PL. Graphical depictions of a conceptual framework for the proposed study are given in Figure 1.

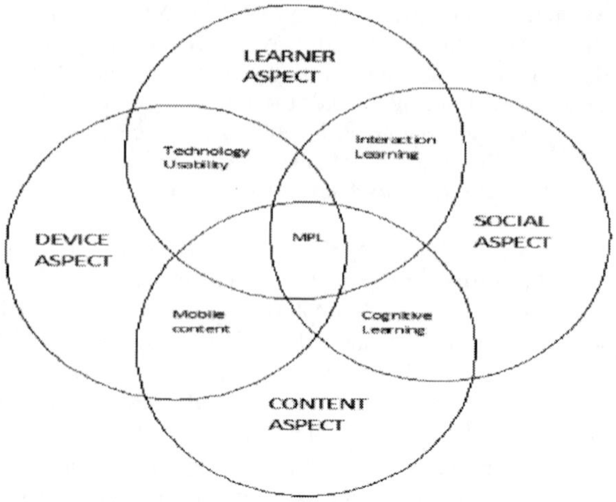

Figure 1. Conceptual Framework.

The intersection between learner and device aspect (technology use) describes the complexity and usability of the technology in terms of accessibility, portability and ergonomics. Whilst the intersection of the device and content (Mobile content) addresses the type of content suitable to run on portable devices and how they need to be customized. The intersection between content and social aspects (Cognitive learning) describes how the learner engages with the content based on their social interactions with their peers and instructors, which may affect how they interpret and construct meaning of their own learning. Lastly, the interaction between learner and social aspects (interaction learning) addresses the collaboration issues and interactive learning that promotes learning pertaining to their unique cultural contexts that may influence how they learn individually and from each other. The study proposes that when all these identified aspects are met and their four dimensions are well balanced, a conducive mobile personalized learning environment can be created.

V. METHODOLOGY

This study used content analysis to search for factors that influence the effective implementations of PL in higher education. Relevant keywords such as "personalization", "learning with technologies", "pedagogical tools for learning", "education", "enhanced learning", "collaboration", "student-centered learning", "ubiquitous learning", "adaptive learning" were used to direct all searches on journal and conferences in the three well know databases JSTOR, Web of Science, and Science Direct. Furthermore, thorough searches were conducted on the following scientific journals which were considered key journals for the field due to their focus on educational use of technology. These include Education and Information Technologies, Journal of Research and Practice in Information Technology, Interactive Technology and Smart Education, British Journal of Educational Technology Computers & Education, Journal of Research on Technology in Education, International Journal of Engineering & Technology, Computers in Human Behavior, International Journal of Education and

Development using Information and Communication Technology, Universal Journal of Educational Research, Journal of Educational Computing Research, Educational Technology Research and Development.

The total searches were limited to the period from 2014 to 2018 due to this pedagogy being at its infant and gaining popularity in the education domain. The results documented in a search protocol were then cleared of any articles that are identified to fall outside the research area of interest-i.e. PL in formal education. 30 articles were read and categorized as relevant based on the preliminary criteria with the intention to identify more themes in the literature. The criterion was inclusive of the following: the novelty of the research, empirical data presence, learning theories applied, model design, technological prototyping, implementation, evaluation and recommendation, and applied educational practices. The criterion was chosen with the intent to help identify literature that deals with different aspects on learning with technology, specifically in PL environments to enhance student performances, and pedagogy. The identification of themes and key factors contributing to the effectiveness of mobile computing PL model were identified following the reading of the articles.

Ms excel was used to tabulate the results and to count the frequency of the appearance of each factor. Factors that were similar in concepts were placed in the same category whereas those with the same meaning were renamed and given a common name. By so doing, many factors were eliminated and the remaining factors were placed into five categories. Table 1 demonstrates the factors tallied with their frequency of appearance in the review literature.

TABLE I. FACTORS INFLUENCING PL

Factors influencing PL effectiveness					
Pedagogy	***Freq***	***Device/ Technology***	***Freq***	***Leaner***	***Freq***
Pedagogical skills	21	Compatibility	23	Experience	11
Learning theories	16	Complexity	18	Education & Training	7
		Flexibility	18	Learner diversity	16
		System quality	16	Learning styles	17
		Internet connectivity	22	Use of technology	16
Content	***Freq***	***Social***	***Freq***		
Content presentation	9	Collaboration	15		
Content development	8	Interactive learning	14		
Information quality	14				

VI. FUTURE WORK AND CONCLUSION

Mobile technologies are considered a new form of social and cultural artifact that mediates teaching and learning in education. Emerging personalized learning spaces along with mobile technologies in the African context aims to better understand how these technologies can enhance quality of learning while promoting accessibility to learning resources. Therefore, it is imperative to look into a holistic approach to education that must be studied in its natural surrounding where personal, situational, cultural and socio-economic factors play a role. PL along with mobile technologies in the African context aims to look into how use of technology can enhance the quality of learning. More so, by extending educational resources to ensure they are readily available to learners anytime and anywhere with no geographical boundaries. This therefore calls for longitudinal studies to be conducted, not just pilot projects that investigate these PL spaces with an attempt to transform pedagogical approaches with authentic contexts that are linked to developing high order thinking.

Mobile computing and technologies have promoted a new vision for learning, particularly in higher education. However, there still exist challenges in education as learning should be fundamentally personal, flexible, social, ubiquitous, complex and dynamic in nature [12]. Therefore, there is a need for a shift in pedagogy towards a more personalized learning environment as opposed to a one-size fits all and passive learning solutions. This requires learning designs that are based on use of mobile technologies that are capable of supporting individual learners to produce desirable learning outcomes. Literature indicates that use of mobile phones have gone beyond making calls and texting due to internet and the Web, which has led web browsing, social media as top uses [11]. The concept of mobile personalized learning aims to put emphasis on the use of mobile technologies and personalization to make learning better in terms of accessibility, portability and mobility, thus eliminating any geographical boundaries. This paper aims to extend flexibility of PL through enhancements of existing models of teaching, and learning with technologies as studies indicates that they are currently minimal [2].

REFERENCES

[1] Z. Abdrakhmanova, K. Aisultavona, L. Chaltikenova, Satkenova, Z. and B. Zhanat, "Implementation of Modern Computer Science and Information Technologies in Teaching," International Business Management. Vol.11, pp. 469–473, 2017.

[2] M. Al-emran, H. M. Elsherif, and K. Shaalan. "Investigating attitudes towards the use of m-Learning in higher education," Computers in Human Behavior. Vol. 56, pp. 93–102. 2016, Doi: 10.1016/j.chb.2015.11.033.

[3] M. Alrasheedi, L.F. Capretz, and A. Raza, "Management's perspective on critical success factors affecting m-Learning in higher education institutions — an empirical study," Journal of Educational Computing, Vol. 54, pp. 253–274, 2016, Doi: 10.1177/0735633115620387.

[4] E. N. Asiimwe, Å. Grönlund, and M. Hatakka, "Postprint Practices and challenges in an emerging m-learning environment," Interntional Journal of Education and Development using Information and Communication Technology, Vol.13, pp. 103–122, 2017.

[5] M. K. Bachman, and C. G. Gannod, "A critical analysis of m-learning initiatives," IADIS International conference, Avila, Spain, pp. 310-312, 2011.

[6] A.J. Bingham, J.F. Pane, E.D. Steiner, and L.S.Hamilton, "Ahead of the Curve: Implementation Challenges in Personalized Learning School Models," Educational Policy,

[7] K.J. Burden, and M. Kearney, ""Investigating and critiquing teacher educators' mobile learning practices", Interactive Technology and Smart Education, 2017, vol. 14, pp. 110–125.

[8] F.V.O. Callaghan, D.L. Neumann, L. Jones, and P.A. Creed, "The use of lecture recordings in higher education : A review of institutional , student , and lecturer issues", Education and Information Technologies," pp. 399–415, 2017, doi: 10.1007/s10639-015-9451-z.

[9] M. J. Casany, M. Alier, E. Mayol, J. Piguillem, and N. Galanis, "Moodbile : A framework to integrate m-learning applications with the LMS," Journal of Research and Practice in Information Technology, Vol. 44, pp. 129–149, 2012.

[10] P.Cerna, "Mobile Computing Framework for Student Engagement System in Ethiopian Higher Educational Institution," International Journal of Wireless Communications and Mobile Computing, 2018, vol. 6, pp. 10–19, doi: 10.11648/j.wcmc.20180601.12.

[11] H. Crompton, "Interactive technology and smart education article information ," Interactive Technology and Smart Education, Vol. 14, pp. 97–109, 2017.

[12] A. Hevner, S.T. March, and J. Park, "Design Science in Information Systems Research," MIS Quarterly. Vol. 28, pp.75–105, 2004

[13] W. Holmes, S. Anastopoulou, H. Schaumburg, and M.Mavrikis, "Technology-enhanced Personalised Learning: Untangling the Evidence," The Robert Bosch Stiftung GmbH, Stuttgart, 2018.

[14] A.K. Hostler, "Framing Mobile Learning: Investigating the Framework for the Rationale Analysis of Mobile Education In Keengwe J. & Maxfield, B.M," Advancing Higher Education with Mobile Learning Technologies: Cases, Trends, and Inquiry-Based Methods, 2015.

[15] A. Juvova, S. Chudy, P. Neumeister, J. Plischke, and J. Kvintova,. "Reflection of Constructivist Theories in Current Educational Practice," Universal Journal of Educational Research. Vol. 3, pp. 345–349, 2015, doi: 10.13189/ujer.2015.030506.M. L. Koole, " A model for framing m-Learning. In M. Ally (Ed.), M-Learning: Transforming the delivery of education and training," Athabasca, AB: Athabasca University Press, 2009.

[16] M. L. Koole, " A model for framing m-Learning. In M. Ally (Ed.), M-Learning: Transforming the delivery of education and training," Athabasca, AB: Athabasca University Press, 2009.

[17] T. Laohajaratsang, "The effect of tablet-based learning on pre-service teachers' learning experiences at Chiang Mai University," IEEE 63rd Annual Conference International Council for Educational Media *(ICEM)*, 2013, pp. 1-11.

[18] C.L. Lai, G.J. Hwang, and J.C. Liang, "preferences of high school teachers and students in Taiwan : a structural equation model analysis," Educational Technology Research and Development. Springer US, Vol.64, pp. 533–554, 2016, doi: 10.1007/s11423-016-9432-y.

[19] L. Lindsay, "ransformation of teacher practice using mobile technology with one-to-one classes: M-learning pedagogical approaches," British Journal of Educational Technology. Vol.45, 883–992. 2016, doi: 10.1111/bjet.12265.

[20] S. Meacham, A. Stefanidis, L. Gritt, and K. Phalp, "Internet of Things for Education: Facilitating Personalised Education from a University's Perspective," Thwenty Third International Conference on Technology in Education, 2018, pp.69–82.

[21] Mercado, and P. Agustin, "Mobile application integration framework for educational institutions," in The Tenth International Conference on eLearning for Knowledge-Based Society, 2013, pp. 7.1-7.10.

[22] S. Olusegun, "Constructivism Learning Theory : A Paradigm for Teaching and Learning," The International Journal of Research & Method in Education, Vol.5, pp.66–70. 2015, doi: 10.9790/7388-05616670.

[23] C. Pimmer, M. Mateescu, and U. Grohbiel, "Computers in Human Behavior Mobile and ubiquitous learning in higher education settings: A systematic review of empirical studies," Computers in Human Behavior, Vol. 63, pp. 490–501, 2016, doi: 10.1016/j.chb.2016.05.057.

[24] A. Phillipson, A. Riel, and A. B. Leger, "Between Knowing and Learning : New Instructors ' Experiences in Active Learning Classrooms," The Canadian Journal for the Scholarshi of Teaching and Learning, Vol. 9, pp. 1–20, 2018.

[25] L. Ross, E. Judson, C.J. Ankeny, S.J. Krause, and R.J. Culbertson, "Is There a Connection Between Classroom Practices and Attitudes Towards Student-Centered Learning in Engineering ?," 2018 ASEE Annual Conference & Exposition, "Is Technology good for education?," Polity Press, Cambridge, 2018, doi:10.1080/00220671.2016.1253948.

[26] S.W. Tabor, "Making Mobile Learning Work : Student Perceptions and Implementation Factors," Journal of Information Technology Education: Innovations in Practice, Vol. 15.pp. 75–98, 2016.

[27] E. Verdú, L.M. Regueras, E. Gal, J.P. de Castro, M.J. Verdú, and D. Kohen-Vacs, "Integration of an intelligent tutoring system in a course of computer network design," Educational Technology Research and Development, vol.65, pp. 653–677, 2017, doi: 10.1007/s11423-016-9503-0.

[28] B., Waldrip, J. Jin, & Y. Vaughan, "Validation of a model of personalized learning," Learning Environments Research, Vol. 19, pp. 169–180, 2016, doi 10.1007/s10984-016-9204-y.

[29] D. Zou, and F.L.Wang, "A flexible graph-based model for facilitating digital learning activities," International Journal of Innovation and Learning, Vol. 23, pp. 444-462, 2018.